2011

INTERNATIONAL DESIGN EXCHANGE PROJECT

URBAN ROOFTOPS

01
LONDON METROPOLITAN University
Sir John Cass Department of Art, Media and Design
Interior Design and Technology BA (Hons)
UK

02
DONGYANG MIRAE University
Department of Interior Design
Rep. of KOREA

The 2nd
Hidden Space Project

URBAN ROOFTOPS

The 2nd Hidden Space Project

Urban Rooftops_35 Projects

c o n t e n t s

02 DONGYANG MIRAE University _Rep. of KOREA

INTRODUCTION

PROJECTS

Tom Jones BArch
(Dist), BSc (Hons), UWCC. RIBA
Associate Principal

POPULOUS
Major projects
London 2012 Olympic Stadium
Sinaport Sports Hub
Emirates Stadium, London
Arsenal FC HQ, London
ICC Cricket World Cup 2007
Antigua stadium, Antigua

Rooftop foreword

The roof of a building is one of the most important elements of both design and construction. It marks the point at which a building touches the sky and defines its silhouette. It is generally the point of contact with the elements and as such has to respond to the needs of shelter and shade, but also offers opportunities for unique vantage points and a fresh perspective on its surroundings.

One of the most significant moments in any building project is the 'topping-out' ceremony, which marks the point at which the highest point of a structure is completed. Mythology has built up about the origins of this ceremony, but the significance of reaching the highest point of any building has been celebrated throughout time in a variety of ways.

Rooftops have often combined functional requirements with artistic design to create wonderful forms. These range from the windtowers that assist with natural ventilation in the middle east, to the clock towers of churches, through to Gaudi's elaborate exploration of chimneys and roof terraces at the Casa Mila house in Barcelona.

However, in recent times, modernism has stripped away the exuberance of rooftops and left them as flat roofs — with little charm or useful function. The top of a building became a simple line that capped-off construction and in many cases then became a dumping ground for mechanical plant and communication aerials. The concept of the rooftop as the 'fifth elevation' was largely ignored and little attention was paid to the roofscape, resulting in an unsightly collection of 'lids' and cooling equipment!

In recent times, the opportunities of using rooftops for both practical and aesthetic purposes has been restored. This has often been as a result of the increasing demands for land in cities and the resulting escalation in land value prices, but also reflects a recognition of the unique qualities of rooftop space. These parameters combine in penthouses, where the desire for return on investment from developers and the desire from customers for rooftop living has led to the creation of a new form of residential space at the top of buildings. The penthouse flat exploits views and opportunities for indoor and outdoor living, while providing a space that attracts a considerable premium in sales value.

New York city has led the way in challenging the perception of rooftops, with a wide variety of spaces that accommodate working, living and playing. Rooftops provide spaces for weddings, dining, spa baths, bars, parks and even agriculture – with a number of urban farms being set up on the large rooftops of industrial units. Asia is responding to this with new buildings that place outdoor leisure space and sports facilities on the roof in new and imaginative ways, taking advantage of the opportunities that rooftops provide.

The aim of this project was to take an existing rooftop and explore new and creative ways of using this space. I hope that the students will be excited by this challenge and I look forward to having my own perceptions of rooftops challenged by their design proposals!

서 문

건물의 옥상은 디자인과 건축 모두에 있어서 가장 중요한 요소들 가운데 하나이다. 옥상은 그 건물이 하늘과 맞닿은 지점이 되며, 건물의 전체적 실루엣을 결정짓는다. 일반적으로 옥상은 다른 요소들과 조화를 이루어야 하며, 외부로부터 건물 내부를 보호하는 역할을 수행하게 된다. 하지만 이와 동시에 옥상은 주변 환경을 새로운 시각에서 조망할 수 있는 특별한 기회를 마련해주기도 한다.

모든 건축 프로젝트에 있어서 가장 의미있는 순간이라면 다름아닌 상량식을 꼽을 수 있을텐데, 건물에서 가장 높은 구조물이 완성되었음을 의미하기 때문이다. 이 행사의 유래에 대해서는 구구한 전설들이 전해지지만, 건축물의 가장 높은 지점에 도달한다는 것은 언제나 중요한 의미를 지니며, 긴 세월 동안 다양한 방식으로 이를 기념해왔다.

옥상을 통해 기능적 필요성과 예술적 디자인이 조화를 이룸으로써 아름다운 구조물이 창조되는 경우가 허다하다. 중동에서 건물의 환기를 돕기 위해 설치하는 윈드타워(windtower)나 교회의 시계탑은 좋은 사례가 될 것이며, 바르셀로나의 까사 밀라(Casa Mila)에서 가우디(Gaudi)가 보여준 굴뚝과 지붕 테라스에 대한 섬세한 탐구 역시 빼놓을 수 없을 것이다. 하지만 근래에는 모더니즘의 영향으로 인해 풍요롭던 옥상의 모습은 사라지고, 매력이나 유용한 기능은 찾아볼 수 없는, 밋밋한 지붕만이 남게 되었다. 건물의 최상부는 구조물이 이제 더 이상 위로 올라가지 않는다는 것을 나타내는 단순한 직선 형태를 띠게 되었고, 대개의 경우 기계 설비나 통신 시설을 대충 설치해 둔 야적장 같은 꼴이 돼버렸다. 옥상에 대한 창의적 컨셉트는 내팽개쳐졌고, 스카이라인에 대한 관심 역시 거의 없기 때문에 하늘은 보기 흉한 건물의 "뚜껑"과 냉각 장치들로 뒤덮혀버렸다.

최근 실용적인 동시에 심미적 차원에서 옥상을 활용할 수 있는 기회들이 다시금 나타나고 있다. 도시에서의 토지 수요 증대와 이에 따른 지가 상승이 그 원인이기도 하지만, 옥상 공간의 독특한 성격에 대한 새로운 인식이 바탕에 깔려있다. 이러한 요인들이 결합되어 등장한 것이 바로 펜트하우스인데, 개발 이익을 회수하려는 개발업자들과 옥상 공간에서 살고싶은 입주자들의 이해관계가 맞아떨어져 건물 꼭대기에 새로운 형태의 주거 공간이 탄생하게 된 것이다. 펜트하우스는 실내외 공간을 모두 활용할 수 있다는 점과 탁월한 전망을 내세워 매매 시 상당한 프리미엄을 갖게 되었다.

뉴욕시는 업무, 주거, 유흥 등의 목적으로 사용되는 다양한 공간을 통해 옥상의 기존 개념에 정면으로 도전하고 있다. 옥상은 결혼식, 다이닝, 스파, 바, 공원, 심지어 농업 등 다양한 용도로 활용되고 있으며, 산업 시설 위에 자리잡은 넓은 옥상 공간에는 수많은 농장이 만들어지고 있다. 아시아 역시 이러한 추세에 발맞춰, 신축 건물들에는 새롭고 창의적인 방법으로 아웃도어 레저 공간과 스포츠 시설물이 옥상에 배치됨으로써, 옥상이 제공하는 기회를 십분 활용하고 있다.

본 프로젝트의 목표는 기존의 옥상 공간을 살펴보고, 이 공간을 활용할 수 있는 새롭고 창의적인 방법을 모색하는 것이다. 학생들이 이러한 도전에 흥미를 갖길 기대하며, 자신들만의 디자인 아이디어를 통해 옥상에 대해 필자가 제시하고 있는 개념에 도전할 수 있기를 기대한다.

LONDON Metropolitan University
Sir John Cass Department of Art, Media and Design
Interior Design and Technology BA (Hons)

Sir John Cass Department of Art, Media and Design
41-47 Commercial Road
Lonodon E1 1LA
Switchboard 020 7423 0000
www.londonmet.ac.uk

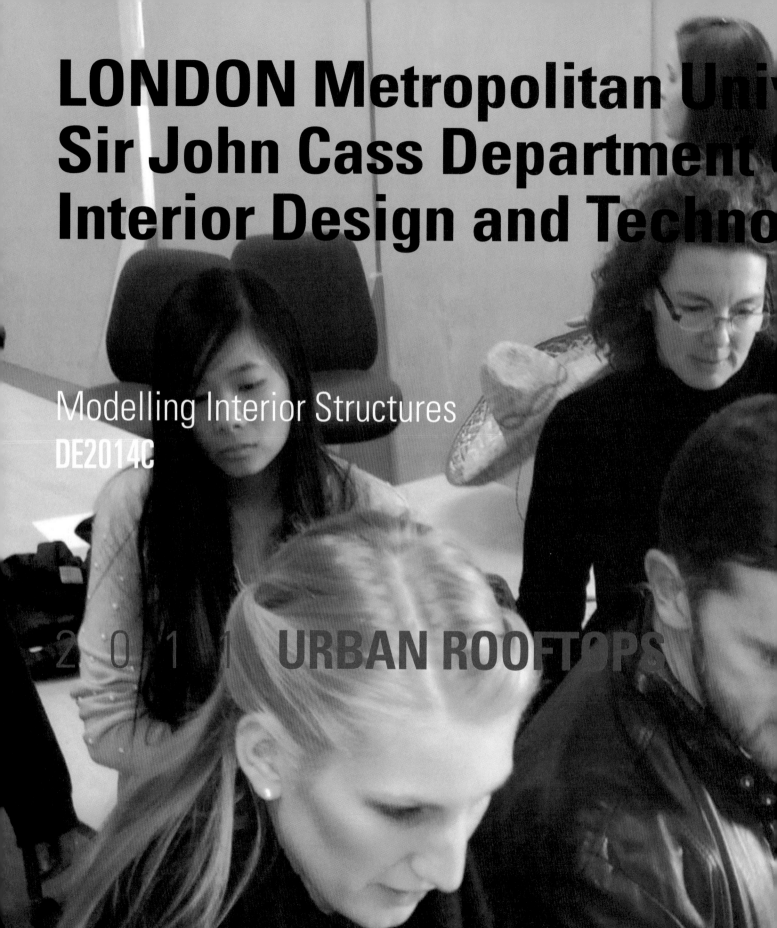

LONDON Metropolitan Univ
Sir John Cass Department
Interior Design and Techno

Modelling Interior Structures
DE2014C

2 0 1 1 URBAN ROOFTOPS

rsity
Art, Media and Design
BA (Hons)

Panorama – **A**

Panorama – **B**

Panorama – **A'**

Panorama – **B'**

Kaye Newman

Course Organizer
Interior Design and Technology BA (Hons)
London Metropolitan University
Sir John Cass Department of Art, Media and Design
Interior Design and Technology BA (Hons)

Hidden Space

Urban Rooftops :

Our cities are now having to rethink and redevelop the use of space more creatively and ingeniously than ever before. Space is precious in any context but the extra pressure cities' face with the ever growing need to accommodate increasing populations mean that every open area is now a new possibility for spatial development.

One of the most recent and fascinating trends is the use of rooftops. Their appeal lies in their surprise and undiscovered disposition. Their purpose might involve a large community or an individual but because of its elevated position, each journey to that roof space provides a chance for adventure. They are an addition to older established buildings but come with a new brief and therefore a new spirit, encouraging creativity and positive thinking, using a different energy.

The master planners and the local communities need to consider a more responsible composition of ideas to our urban and earthly needs. Sustainable energy use and material choice are to be rewarded but what of the function? Is just viewing the simple horizon above the rooftops function enough? relieving us of our daily stresses and seeing the city as sleeping giant or should we be industrious, striking balances of nature, pulling it back from disaster, giving space to synergetic projects.

The students were asked to design a Rooftop structure in a densely inhabited urban neighbourhood, namely Burbage House, Curtain Road, Shoreditch London E2.

Aim

The aim of this lofty structure or series of structures sets out to encourage, inspire a community or an individual to cultivate positive practices.

The rooftop should connect and bond visually or through metaphor within its surrounding neighbourhood. The space should take account of the social and cultural diversity that resides close by.

Context

Shoreditch is weave of close knit streets with its former industrial context provide an area for those that want to step out of the norm and the corporate, it's place to imagine new business. Innovative creative houses have set up home using space imaginatively, sharing with like minded disciplines.

The narrowness of the streets encourages networking and enhances collective working practices. Warehouse constructions reveal their previous activity and function, reminding us of the busy people, the noise and the industry.

The buildings allow a maximum of light though to the interior, the detail in the window frames and brickwork show a sense of pride. The Interiors are simple and true the outer structure, revealing angles and curves.

The space is unattractive to the Multi-national corporations, and as there is little capacity for charging large rents makes this area economically suitable for young and emerging practices to set up.

The rawness of the interior spaces, unadulterated by recent 20th century trends provide the perfect utilitarian and vacant habitat for the fresh ideas where pioneering and ground breaking professions materialize.

View

In some respects the rawness of such a landscape, its lack of conformity compares well with an open landscape. When standing on the roof looking across London, its openness and horizontal neutrality engenders a sense of calm and invigoration. Suddenly there are possibilities and scope for living positively. Hurried life slows down and London becomes picturesque.

Novalis wrote 'Everything seen from a distance becomes poetry :distant mountains, distant people, distant events. Everything become romantic'.

숨겨진 공간

Kaye Newman
Course Organizer - 코스 기획자

도시의 옥상

현대 도시들은 그 어느 때보다 더욱 창의적이고 독창적으로 공간 활용과 개발에 대해 고민할 필요가 있다. 공간은 어떤 환경하에서든 소중하지만, 도시가 증가하는 인구를 감당해야 할 필요성이 점증하는 현 시점에서, 모든 열린 공간은 공간 개발을 위한 새로운 가능성을 품고 있다.

옥상의 활용은 최근 나타난 가장 흥미로운 트렌드 가운데 하나이다. 옥상의 매력은 아무래도 새롭고 놀라운 방식의 공간 배치일 것이다. 옥상의 용도는 다양할 수 있다. 하지만 한 가지 분명한 사실은 높은 곳에 위치하고 있는 옥상에 올라가는 것은 일종의 모험이 될 수 있다는 것이다. 옥상 공간이 낡고 오래된 건물에 추가로 만들어지더라도, 새로운 감각을 수반하기 마련이며, 이로 인해 색다른 에너지를 바탕으로한 창의성과 적극적 사고의 가능성이 증대된다.

도시계획입안자들과 지역사회는 도시의 요구에 부응함에 있어 보다 책임감 있는 아이디어를 강구해야 한다. 지속가능한 에너지 및 소재의 선택과 사용은 칭찬받아 마땅하지만, 기능성에 대해서는 어떤 고민을 하고 있는가? 하루의 스트레스를 풀기 위해 그저 옥상 위로 펼쳐진 스카이라인을 바라보는 것만으로 만족할 것인가? 아니면 보다 부지런한 자세로 이러한 공간에 에너지를 불어넣을 새로운 프로젝트를 계획할 것인가?

학생들에게 주어진 과제는 인구밀도가 높은 도심지인 런던의 쇼디치 지역 커튼 로드에 위치한 버비지 하우스, Curtain Road, Shoreditch, London의 옥상을 디자인하는 것이다.

목표

프로젝트의 대상이 된 구조물이 애초에 의도했던 목표는 커뮤니티 또는 각 개인들이 보다 적극적인 활동에 나서도록 고무하는 것이다. 옥상은 시각적으로나 상징적으로 주변 지역을 연결하고 이어주어야 하며, 동시에 근방에 자리잡고 있는 사회적, 문화적 다양성을 반영할 수 있어야 한다.

배 경

쇼어디치는 촘촘히 짜여진 도로망을 지닌 과거 공업 지역으로서, 일반적인 기준을 벗어나 새로운 형태의 사업을 꿈꾸는 사람들에게 기회의 땅이 되고 있다. 뜻을 같이 하는 사람들이 공간을 창의적으로 활용한 혁신적이고, 창조적인 주택들을 이곳에 짓고 있다.

좁은 도로로 인해 사람들 사이의 연대가 강화되고, 공동작업은 활기를 띄게 된다. 창고 건물들은 우리에게 바쁘게 일하는 사람들, 소음, 그리고 공장의 모습을 다시금 떠올리게 한다. 건물들은 최대한의 채광을 고려해 지어졌고, 창틀과 벽돌에서는 일종의 자부심이 배어 나온다. 인테리어는 단순하고 외부구조와 조화를 이루며, 모서리와 만곡은 그대로 드러난다. 이 공간은 다국적 기업 등에게는 그닥 매력적이지 않을 뿐더러, 높은 임대료를 책정할 여지도 거의 없기 때문에 패기만만한 신규사업자들에게 적합하다 할 수 있다.

최근의 경향이 전혀 반영되지 않은, 날것 같은 느낌의 내부공간은 진취적이고 혁신적인 사업이 뿌리내릴 수 있는 완벽하고도 실용적인 장소를 마련해준다.

조 망

건물이 갖고 있는 날것의 느낌, 그리고 기존의 질서에 편입되지 않은 듯한 특성들은 탁트인 경관과 훌륭한 조화를 이룬다. 지붕에 올라 런던 시내를 둘러보면, 그 개방성으로 인해 평온함과 에너지를 동시에 느낄 수 있다. 새로운 가능성과 적극적 삶에 대한 희망을 동시에 느끼게 해준다. 바쁘게 살아온 삶의 속도는 느려지고, 런던은 어느새 그럴싸한 풍경으로 다가온다. 독일 시인 노발리스(Novalis)는 이렇게 노래했다.

"멀리 떨어져서 바라보면 모든 것이 시가 된다네. 멀리 있는 산들, 멀리 있는 사람들, 그리고 멀리 있는 사건들 모든 것이 로맨틱해진다네."

Janette Harris

Senior Lecturer
Module Leader
Interior Design and Technology BA (Hons)
London Metropolitan University
Sir John Cass Department of Art, Media and Design
Interior Design and Technology BA (Hons)

Modelling Interior Structures
DE2014C

Module Summary

The module is concerned with the exploration and realisation of structure and materials within a space. Workshop and class based activities encourage a wide range of skills, highlighting sensitivity to design matters, reflective practice, informal debate, peer assessment and presentation. Software is used to produce orthographic details, visuals and drawings for the production of components. Scale is explored in the context of human interaction. Spatial concepts are considered through modelling in three dimensions, using a wide range of machinery and hand tools, allowing the student to fully explore all modelling possibilities. Process is developed using proven and experimental techniques. The module promotes the growth of existing skills combined with the exploration of the potential of technology.

The Pedagogy

The twelve week designate module (one of four) runs in the first semester of the second year. Students, by this stage, have a good understanding of 3D studio Max, Photo shop and Autocad, however the initial emphasis is on drawing, sketch modelling and photography, in order to explore concepts through experimentation. This experiential approach develops reflective practice, by challenging student's original ideas, opening up opportunities for discussion and debate. Activities are distributed across the taught sessions of the module to engage and broaden student's knowledge.

The roof top project posed a number of challenges;

- Connecting the site to the context and those who work within the building
- Understanding the aspect of the site through environmental analysis
- Understanding the demographic of the area through analysis and diagnostic
- Understanding the materiality of the existing building and it's surroundings
- Understanding the needs of those working within the confines of the building through a client briefing

The students were given six weeks to research and develop a concept. Our mentor on the project Tom Jones from Populous, critiqued the concepts at this point offering guidance for the continuation of the projects. From this point the students started sketch modelling. By week ten all the students were either starting a professional model in appropriate materials or drawing the project in 3D max for rapid prototyping.

Constructivist View of Learning

The students are taught using constructivist learning theories that develop a learning community through communication and cooperation through activity, based on work by Vygostsky 1978 influenced Lave & Wenger 1991 idea of Learning Communities

Activity

develops knowledge construction

Learning is developed by

- building on existing knowledge
- students being given sufficient time to generate concepts, through a period of incubation
- tasks were provided to keep the students within a timeframe of learning
- developing skills through experimentation and review

Activities of construction understanding

- Interaction with material systems and concepts in the domain
- Interactions in which learners discuss their developing understanding and competence.

Design Principles for constructivist teaching and learning activities

- Ownership of the task
- Coaching and modelling of thinking skills
- Scaffolding
- Guided discovery
- Opportunity for reflection
- Ill-structured problems

Supporting the students face-to-face sessions is an online learning environment called WebLearn, part of Blackboard. The e-learning environment guides and develops students learning through week by week tasks, documentation, links and images, encouraging the following:

- Discussion and debate through a blog
- Students encouraged to develop a resource library for the module for the whole class to use
- Students set up study groups to share practice and process developing a community of practice
- Students to develop reflective practice by using an online portfolio
- Feedback given to students through mail within WebLearn and writing their personal portfolio's
- Feedback from the students to the tutor through an anonymous online proforma
- Peer and Self assessment.

Students are supported through out with small group and one-to-one tutorials, building their confidence to discuss and debate their design intentions. This reflects in the individuality of the outcomes demonstrated in the sketchbooks and final outcomes.

The collaboration with Kyungwon University and DongYang University in Seoul South Korea, has given the students opportunity to demonstrate their perspective on the Hidden Space Rooftop project, and this book offers us an insight to three Universities approaches. We are very proud to be part of this project.

Janette Harris
Senior Lecturer
Module Leader – 교과목 지도교수

모듈(교육과정)개요

본 모듈은 어떤 공간 속에서 구조물과 재료들을 탐색하고 구체화시키는 데 역점을 두고 있다. 워크샵과 수업시간 동안의 활동들을 통해 다양한 기술들을 익힐 수 있는데, 디자인 소재에 대한 감수성, 격의 없는 토론, 동료 학생에 대한 평가, 그리고 발표 등이 대표적이다. 시각이미지와 드로잉 등을 처리할 수 있는 소프트웨어가 활용된다. 사람들 사이의 상호작용을 최대한 고려하며, 공간 컨셉트는 3차원 모델링을 통해 이루어진다. 다양한 기계 및 수공구를 활용하며, 이를 통해 학생들은 모델링과 관련된 다양한 가능성을 탐색할 수 있다. 기존의 기술들 뿐만 아니라 실험적인 기술들도 차용된다. 본 모듈을 통해 기존 기술들의 향상을 도모하는 동시에 기술의 새로운 가능성을 탐색하고자 한다.

페다고지(수업방법)

12주로 이뤄진 본 모듈은 2학년 1학기에 진행된다. 이 즈음이 되면 학생들은 3D 스튜디오 맥스, 포토샵, 그리고 오토캐드 등에 대한 충분한 지식을 갖게 되지만, 수업 초기에는 주로 드로잉, 스케치 모델링, 그리고 사진촬영 등에 주안점을 둠으로써, 학생들이 시행착오를 통해 다양한 컨셉트를 탐구할 수 있도록 한다. 이러한 접근법을 통해 학생들의 기존 생각을 되짚어보는 한편 토론과 논의의 장을 마련함으로써 반성적 사고와 활동을 발전시킬 수 있다. 모듈 내의 각 수업별로 활동을 분산배치함으로써 학생들이 지식의 폭을 넓혀나갈 수 있도록 한다.

학생들에게는 컨셉트를 연구하고 개발할 수 있는 6주의 시간이 주어진다. 본 프로젝트의 멘토인 포뮬러스출신의 톰 존스가 학생들의 컨셉트를 평가하는 동시에 프로젝트를 계속 진행해나갈지에 대한 조언을 제공해주었다. 이 시점부터 학생들은 스케치 모델링을 시작한다. 10주차가 되면 모든 학생들은 적절한 재료로 전문적인 모델링을 시작하거나, 신속한 프로토타입 제작을 위해 3D 맥스를 이용 프로젝트를 드로잉하게 된다.

프로젝트가 진행되는 동안 학생들은 소그룹 또는 1대1 지도를 통해 도움을 받게 되는데, 이를 통해 자신의 디자인 의도를 토론, 논의할 수 있는 자신감을 기르게 된다. 스케치북이나 최종결과물에 나타난 개성을 통해 이 점을 충분히 파악할 수 있다.

한국의 경원대학교 및 동양대학교와의 협력 덕분에 학생들은 히든 스페이스 루프탑(Hidden Space Rooftop) 프로젝트를 통해 자신의 의견을 개진할 수 있는 기회를 갖게 되었으며, 이 책을 통해 우리는 3개 대학의 각기 다른 접근방법에 대한 통찰력을 얻을 수 있었다. 우리는 이 프로젝트에 참여할 수 있었던 것을 매우 자랑스럽게 생각한다.

Alena Netukova
alenanet@live.co.uk

Esprit de Corps

Esprit de Corps

All forms of social identities of communities such as local business and activities (from design, art, the fashion, film to web design, architecture and finance) play a great role in presenting of human beings` life. Soul, intelligence and emotions are reflections of the community. In this context, the project is reflecting the history, the culture and the modern life in the area and the site itself. The aim has been to create a powerful structure defining the synthesis of traditional and modern approaches. Besides its fundamental function, the interior space creates a contemporary level in connection between the public and designers/ artists.

단결심

커뮤니티가 갖고 있는 모든 형태의 사회적 정체성들, 예를 들어 디자인, 미술, 영화, 웹디자인, 건축과 금융 등은 모두 인간의 삶을 드러내는 데 중요한 역할을 담당한다. 영혼, 지적 능력, 그리고 감정에는 커뮤니티의 모습이 투영된다. 이러한 맥락에서 본 프로젝트는 해당 지역의 역사, 문화, 그리고 현대의 삶을 반영하고 있다. 프로젝트의 목표는 전통적인 접근법과 현대적 접근법이 결합된 강력한 구조물을 창조하는 것이다. 내부 공간은, 그 본질적 기능과 더불어, 대중과 디자이너/아티스트를 연결시켜주는 현대적 요구를 충족시켜준다.

Esprit de Corps is to become an identity for the Shoreditch area as well as for the client.

The site ... Burbage House

Located in the heart of the Curtain Road, EC2 the site is occupied by various businesses such as web designers, film, product design, fashion, postal consultants and others.

The project is situated on the roof top of the Burbage house. It serves people from the local businesses as a meeting place with their clients as well as place to have a lunch, BBQ or party.

According to a client`s wishes to accommodate conference, exhibition and party activities, the roof top has been divided into zones partially covered and partially opened. Also a new access has been created by erecting a lift from the main void and that way open the roof top to a public.

The area is characteristic by the schizophrenic traditional `red and yellow London brick`; large warehouse windows reflecting images of close by rich business city.

There is a strong sense of art, fashion and design including expanding small businesses in the whole neighbourhood. From the small cafes, clothes shops, shops offering an exotic cuisine to a fashion design school.

부지 – 버버지 하우스

커튼 로드의 중심부에 위치하고 있기 때문에, 해당 지역에는 웹디자인, 영화, 상품디자인, 패션 등 다양한 직종이 혼재하고 있다.

프로젝트는 버비지 하우스의 옥상에서 진행된다. 이곳은 점심식사나 바베큐, 또는 파티를 여는 등의 목적으로 활용되는 한편, 지역 업체 사람들이 고객을 만나는 장소로도 이용된다.

컨퍼런스, 전시, 파티 등의 행사에 활용하고 싶다는 의뢰인의 바람에 부응하기 위해 옥상은 부분적으로는 개방형으로, 또 부분적으로는 지붕이 덮힌 형태로 구획이 나뉘어졌다. 중앙 공동 구역에 승강기를 추가로 설치함으로써 옥상을 대중에게 개방하는 효과를 얻을 수 있다.

이 지역의 특징은 런던 전통의 적색과 황색 벽돌이 어지럽게 자리잡고 있다는 것이다. 창고의 커다란 창문을 통해 부유한 상업 도시의 이미지를 엿볼 수 있다.

예술과 패션, 그리고 디자인에 대한 강렬한 감수성이 자리잡고 있으며, 이러한 경향은 지역 전역에 걸쳐 소규모 사업

체들로 그 영역을 확장한다. 작은 까페, 옷가게, 독특한 요리를 선보이는 레스토랑에서부터 패션 디자인 학교까지 그 종류는 매우 다양하다.

Graffiti dominate the streets; for some symbol of modern art and expression of artists` feeling towards the outer world and global issues; for others pure act of vandalism. The site is located east from the Tower Wall ... `Jack the Ripper`s London`, in the past covered by dirt and poverty kept away from `clean` imperial London; the difference that can be seen in the architecture. Since, many warehouses have been re-developed and turned into desirable spaces of working and living in. Being so close to the Liverpool Street, Bank and the City, there is a high demand from young professionals, investors and property developers.

These is also the transport factor – time is money – the area is providing fast connection with the city, suburbs and also national connections ... all within a reach.

그래피티가 도로마다 가득하다. 어떤 이들에게는 현대 예술의 상징이자 외부 세계와 범지구적 이슈에 대한 아티스트의 표현으로 받아들여지고, 또 다른 이들에게는 순전히 반달리즘에 불과할 지도 모른다. 부지는 타워월의 동쪽에 자리잡고 있는

데, 말쑥한 런던의 모습과는 사뭇 다른, 먼지와 가난의 흔적이 느껴지는 곳이다. 이러한 차이는 건축물들을 통해 확인할 수 있다. 최근들어 많은 창고들이 재개발되어 업무나 주거에 적당한 공간으로 탈바꿈되었다. 리버풀 스트리트나 시티 지역과 인접해있기 때문에 젊은 전문직종사자, 투자전문가, 부동산 개발업자들 사이에 수요가 매우 높은 편이다.

시간이 곧 돈이란 말이 있듯이, 교통 역시 고려사항이 될 것이다. 이 지역은 시티 및 교외 지역과 빠르게 연결되며, 전국 어디로든 쉽게 이동할 수 있는 교통의 편리성을 지니고 있다.

tracts thoughts of similar nature. The more we consciously think a thought, the easier will be to think that thought again because the connection between the cells will become wider and stronger. The client/ visitors should experience this feeling of bonding and connecting by entering the space — each synapse projects soothing light when active (activated by human body temperature) and the light beam shoots from one synapse to another one and back activating another synapse and creating more light beams and so on... till the whole space is acting like one `smart` brain. Visitors can enjoy a light show in relaxing environment, by grouping in small numbers between the synapses. Or being entertained, lectured in front of 9m high theatre like screen, where the client`s work can be presented; or form a large group in the open air part of the space. Synapses join the ground forming and providing seating areas. Also this can be seen from the distance, and again as a pride of the Burbage house, the Esprit de Corps shines above the whole Shoreditch area.

컨셉트

전체 구조물을 통해 버비지 하우스와 공동체의 구성원들을 한데 묶어주는 자부심과 소속감의 표현을 한 데 아우르고자 한다.

공동체의 결속력은 구성원들이 서로에 대해 갖고 있는 긍정적 태도에서 비롯된다. 구성원들 사이의 상호작용과 의사소통, 공통의 목표와 관심사, 우정, 그리고 외부 세계에 맞서 서로를 지켜주는 힘 등이 공동체의 활동에 영향을 준다. 구조물의 외부는 부분적으로 투명한 곡선 형태로 만들어져 내부공간에 충분한 자연광을 공급할 수 있도록 한다. 이러한 곡선 형태는 시냅스를 이용해 옥상의 내부까지 이어지는데, 이를 통해 인류의 화합을 나타내는 동시에 아티스트/디자이너의 영혼을 표현한다. 인테리어는 마치 인간의 두뇌 속 같은 모습으로 꾸며진다. 모든 생각은 비슷한 성격의 다른 생각들을 불러일으킨다. 우리가 의식적으로 한 가지 생각에 천착할수록, 그 생각을 다시 떠올리는 것은 매우 쉬워지기 마련인데, 세포들 사이의 연결이 보다 광범위하고 강력해지기 때문이다. 클라이언트/방문객도 이 공간에 들어서게 되면 이러한 연결과 연대의 감정을 경험하게 된다. 인간의 체온에 의해 활성화되는 시냅스는 부드러운 빛을 비추게 되고, 이 빛이 또 다른 시냅스를 활성화시키게 되며, 이 과정이 반복되다 보면 어느새 전체 공간이 마치 하나의 "스마트한" 두뇌처럼 빛을 발하게 된다. 방문객들은 시냅스 사이에 삼삼오오 모여 편안한 분위기 속에 빛이 만들어내는 쇼를 즐기게 된다. 혹은 의뢰인의 작품이 상영되는 9미터 높이의 극장형 스크린 앞에서 즐거운 한 때를 보낼 수도 있고, 대규모 모임이라면 옥상의 개방형 공간을 활용해도 좋다. 시냅스는 바닥과 연결돼 앉을 수 있는 공간을 마련해준다. 멀리에서도 확연히 눈에 들어오는 이런 모습은 버비지 하우스의 자부심으로 느껴지며, Esprit de Corps는 쇼어디치 전역에 아름다운 빛을 뿜어낸다.

Concept

The whole structure is a combination of the Burbage house theatrical routes and expression of feelings of the pride and loyalty uniting the members of a group.

As a synergy, the group cohesion formed from the strength of mutual positive attitudes among members of the group. Its functioning is affected by interaction and communication between members, common goals, interests, friendship and support against outside world. The exterior is formed as a fluid shield which is partially transparent to give the interior space plenty of natural light. The fluidity continues into the interior of the roof top by descending the synapses look like elements, which represents the unity of the humans and expresses the soul of the artists/ designer. The interior is like being inside of the human brain; every thought at-

Material Concept

The use of sustainable material is as important for this project as its environmental side.

The structure uses combination of the bio-glass panels and the bio-luminum frames. The bio-glass is 100% recycled glass and is 100%.

The structure uses combination of the bio-glass panels and the bio-luminum frames. The bio-glass is 100% recycled glass and is 100% approach to maximise the thermal absorbance and to balance

between reflectivity and materiality...the effect that neutralizes the heat in the summer and maximiizes the warmth for winters. The Bio-luminum frames are made from the salvaged parts of retired airplanes and as a high-strength they and maximi-

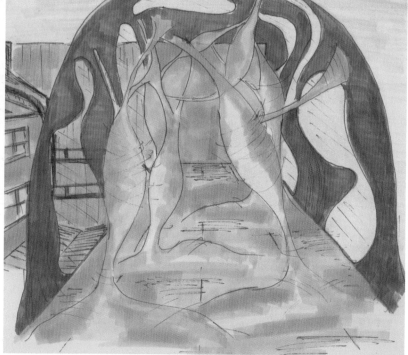

izes the warmth for winters. The Bio-luminum frames are made from the salvaged parts of retired airplanes and as a high-strength they.

자재 컨셉트

이번 프로젝트에 있어서 지속가능한 자재를 사용하는 것은 그 환경적 측면 만큼이나 중요한 사안이다. 구조물에는 바이오유리 패널과 바이오알루미늄 프레임을 사용한다. 바이오유리는 100% 재활용됐을 뿐만 아니라 100% 재활용이 가능한 소재이며, USGBC 프로젝트 인증을 위한 LEED 기준에 부합

한다. 소재로서 바이오유리는 단열 기능을 극대화하는 동시에 반사성과 물성 사이의 균형을 유지하는 것을 목표로 고안되었으며, 이를 통해 여름에는 외부로부터의 열기를 차단하고, 겨울에는 온기를 극대화하는 효과를 얻고 있다. 바이오알루미늄 프레임은 퇴역한 항공기의 부품을 가공해 만들어지는데, 고강도 소재로서 내구성이 매우 뛰어나다. 인테리어에는 흰색의 반투명 코리언(corian)이 바이오 알루미늄의 외부를 덮기 위해 사용되었다.

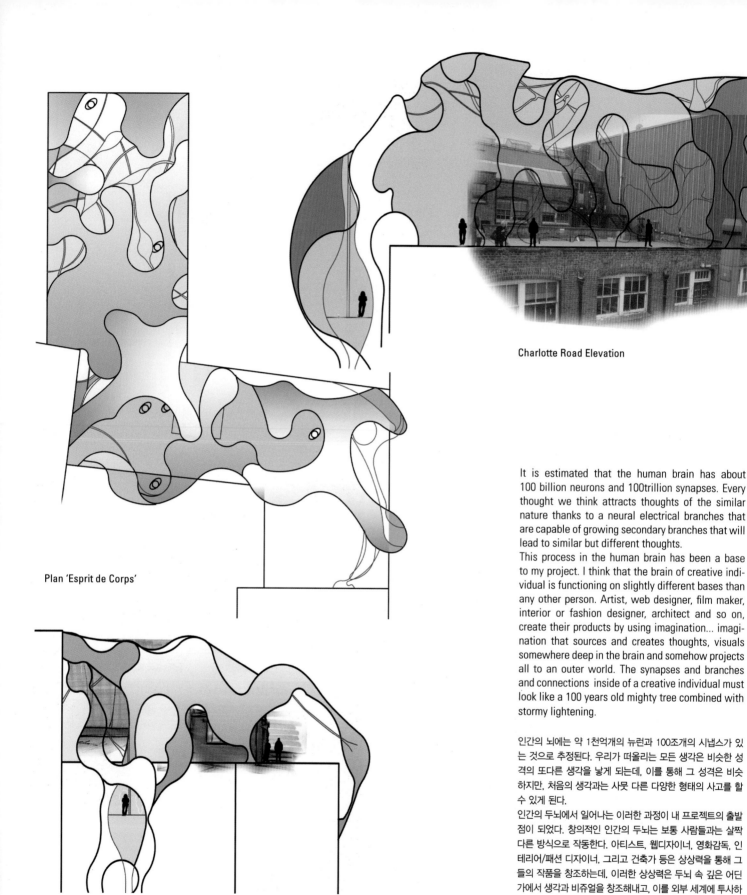

Plan 'Esprit de Corps'

Charlotte Road Elevation

Elevation
showing the rooftop structure extend down to Ground Floor

It is estimated that the human brain has about 100 billion neurons and 100trillion synapses. Every thought we think attracts thoughts of the similar nature thanks to a neural electrical branches that are capable of growing secondary branches that will lead to similar but different thoughts.

This process in the human brain has been a base to my project. I think that the brain of creative individual is functioning on slightly different bases than any other person. Artist, web designer, film maker, interior or fashion designer, architect and so on, create their products by using imagination... imagination that sources and creates thoughts, visuals somewhere deep in the brain and somehow projects all to an outer world. The synapses and branches and connections inside of a creative individual must look like a 100 years old mighty tree combined with stormy lightening.

인간의 뇌에는 약 1천억개의 뉴런과 100조개의 시냅스가 있는 것으로 추정된다. 우리가 떠올리는 모든 생각은 비슷한 성격의 또다른 생각을 낳게 되는데, 이를 통해 그 성격은 비슷하지만, 처음의 생각과는 사뭇 다른 다양한 형태의 사고를 할 수 있게 된다.

인간의 두뇌에서 일어나는 이러한 과정이 내 프로젝트의 출발점이 되었다. 창의적인 인간의 두뇌는 보통 사람들과는 살짝 다른 방식으로 작동한다. 아티스트, 웹디자이너, 영화감독, 인테리어/패션 디자이너, 그리고 건축가 등은 상상력을 통해 그들의 작품을 창조해내는데, 이러한 상상력은 두뇌 속 깊은 어딘가에서 생각과 비쥬얼을 창조해내고, 이를 외부 세계에 투사하게 한다. 창의적인 인간의 시냅스와 신경줄기들은 마치 100년 된 고목과 강렬한 번개가 한 데 어우러진 듯한 모습일 것이다.

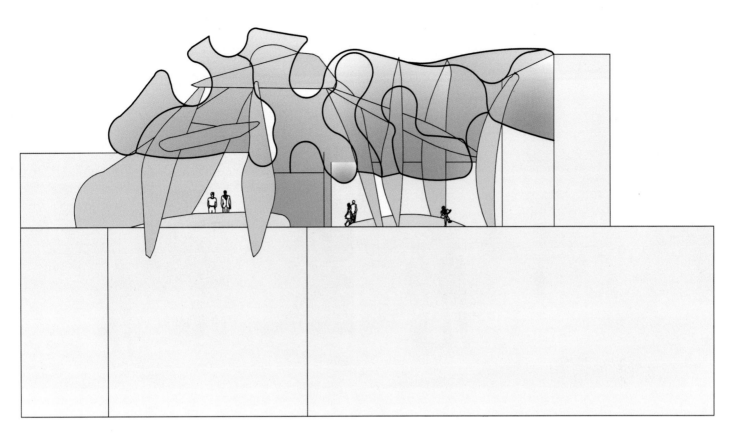

The first and main idea was to create a structure that would represent the connection; the connection between the creative people occupying the surrounding premises, their connection with the public as well as a connection of their artistic and creative brain processes, expressing their soul and emotions and reflection on the Shoreditch area. The roof top would be a space of such a process, reflection and mainly space for the debate.

처음으로 떠올렸고 가장 중요하게 여겼던 아이디어는 연결을 보여줄 수 있는 구조물을 만드는 것이었다. 주변 지역에 살고 있는 창의적인 사람들 사이의 연결, 그들과 보통사람들 사이의 연결, 그리고 그들과 그들의 창조적이며 예술적인 두뇌 활동과의 연결을 꾀했다. 이를 통해 그들의 영혼과 감성, 그리고 쇼어디치 지역에 대한 생각을 표현하고자 했다. 옥상은 이러한 일련의 과정과 사색의 공간이 될 것이며, 무엇보다 토론의 장이 될 것이다.

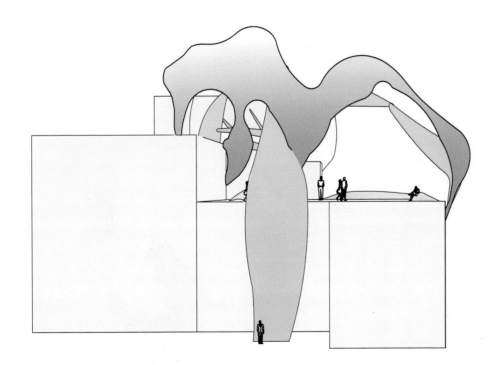

Open and Expansive: the sapce becomes what you want to be.

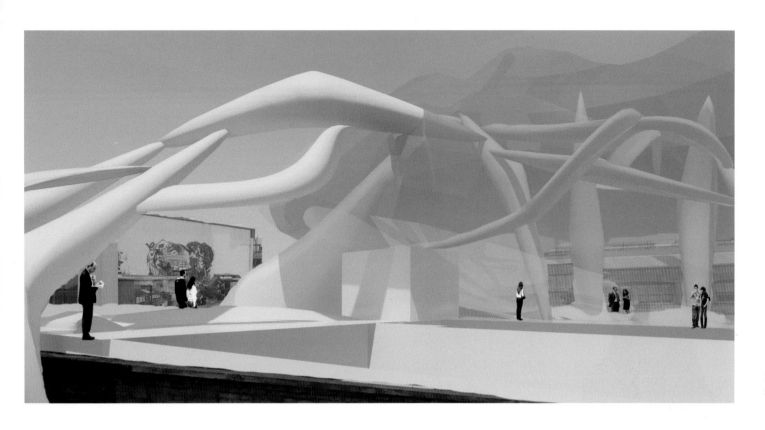

Take time and space to discuss.

Beata Piotrzkowska
bp112358@gmail.com
bdesignportfolio.carbonmade.com

Edge of Time

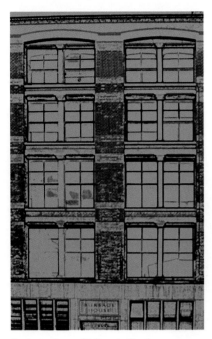

Burbage House 83 Curtain Road

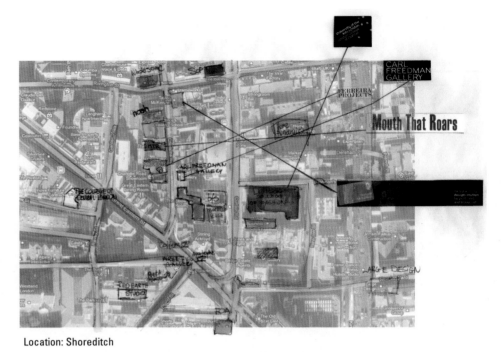

Location: Shoreditch

THE BRIEF

Our cities are now having to rethink and redevelop the use of space more creatively and ingeniously than ever before.
Space is precious in any context but the extra pressure cities' face with the ever growing need to accommodate increasing populations mean that every open area is now a new possibility for spatial development.

Photo montage looking at the free spirit that exists within the shoreditch area

Burbage House

Photographic diagnostic of the rooftop

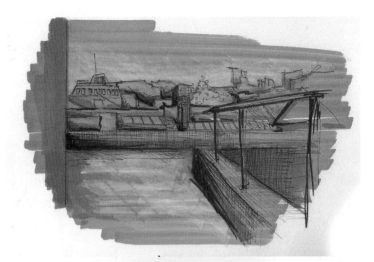

The view from the rooftop is very restricted by the surrounding buildings buildings which are mainly made from bricks; this gives the view a very industrial feeling where everything looks similar. The roof top space is designated for employees.
It is an exclusive, private space which is occasionally made available to friends and family.

주변의 건물들이 옥상에서의 전망을 가로막고 있다. 이 건물들은 주로 벽돌로 만들어져 있고, 이로 인해 모든 것이 비슷하게 보이는, 마치 공장지대 같은 느낌을 자아낸다. 옥상은 직원들에게 할당된 공간이다.
배타적이며 사적인 공간으로서 친구나 가족들에게 간혹 개방되기도 한다.

Juxtapose buildings and brick walls

The shadow lines are recorded as they move across the rooftop.

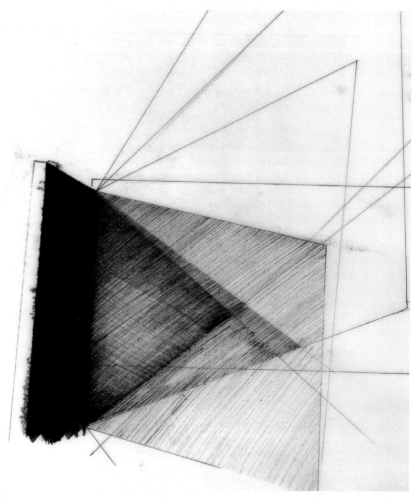

By working with the aspects of light and shadow, changing through the space created a pattern based on the lines of those shadows.

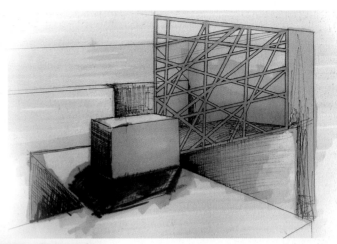

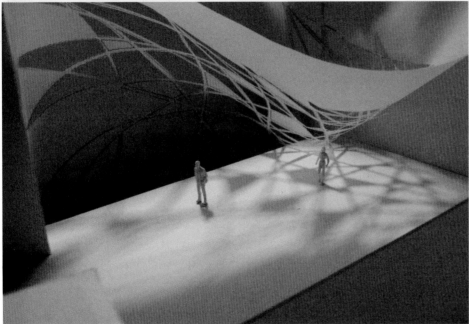

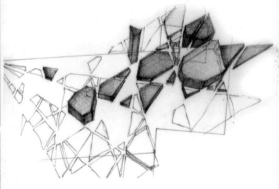

Projecting the light through the pattern creates a calm and relaxing environment which could be constantly changing as the sun moves throughout the day. I created the whole construction which fits above the rooftop by extending some of the lines from the pattern.

공간 위에 생겨나는 빛과 그림자는 시시각각 변하면서 다양한 패턴을 만들어낸다. 빛과 그림자를 통해 만들어지는 패턴은, 태양이 하룻동안 계속 움직이기 때문에, 그에 따라 끊임없이 변화하면서, 고요하고 평온한 분위기를 만든다. 나는 이러한 패턴으로부터 만들어진 선들을 길게 연장함으로써 옥상 위에 딱맞는 건축물을 창조해냈다.

The space will be partly closed to provide shelter for visitors on rainy days. I also used the pattern as a wall from the street view to give an impact from the street level. For the whole construction I want to use white corian to emphasize the contrast with the buildings around and also to calm down the surrounding environment. I have also decided to create stairs in the middle of the space for two reasons. First was the restricted view - I created a new viewing level by going three meters up from the rooftop. The second was the small building which could not be removed so I covered it with the stairs. Traditional square stairs were simply to "heavy" for this space and were definitely overwhelming. I decided to create stairs from different shaped blocks based on the same lines as the pattern I created earlier.

공간의 일부에는 지붕을 덮어 비오는 날에도 방문객들에게 비를 피할 수 있도록 할 것이다. 나는 건축물 전체에 흰색 코리언을 사용함으로써 주변 건물들과의 대조적 느낌을 확보하는 동시에 차분한 환경을 조성하고자 한다. 나는 두 가지 이유 때문에 공간의 중앙부에 계단을 설치하기로 결정했다. 첫번째 이유는 제한된 전망이다. 나는 옥상보다 3미터 높은 곳에 조망을 위한 새로운 장소를 마련했다.
두번째 이유는 건물의 규모가 작다는 점인데, 계단을 이용해 이 부분을 보완했다. 전통적인 사각형 모양의 계단을 설치하면 너무 무거운 느낌이 들 것이며, 아무래도 부담스러운 것이 사실이다. 나는 앞서 내가 만든 패턴의 선들에 맞춰, 각기 다른 모양의 블록으로 계단을 만들고자 한다.

PLAN VIEW Looking at the roof structuire and how its lines map the shadows

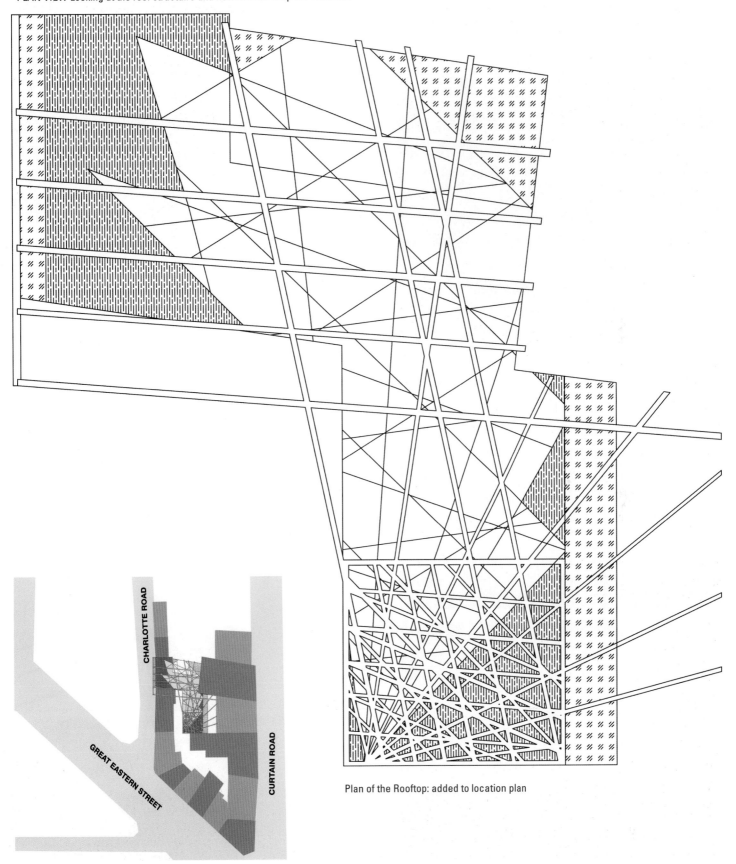

CHARLOTTE ROAD

GREAT EASTERN STREET

CURTAIN ROAD

Plan of the Rooftop: added to location plan

showing the structure cantilvered over the fascia provide a sun baffle.

The aim is to create a space which would change the atmosphere and the feeling of an otherwise ordinary place.

Elevation show street facade cantilvered sru

Rooftop structure looking fro the back of Burbage Hose

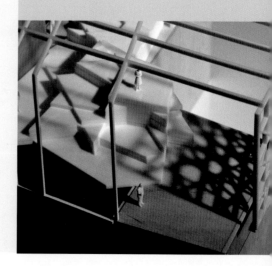

Rooftop facing Charlotte Road

I have decided to create a space which would change the atmosphere and the feeling of an otherwise ordinary place. My first step was to change the proportion of the rooftop by adding the glass platforms along the entrance wall, along the view wall and from the street view. These platforms not only extend the size of the place but by using transparent glass I wanted to create a different feeling and provide a more interesting perspective and view.

프로젝트의 목표는 평범한 장소의 분위기와 느낌을 바꿔놓을 수 있는 공간을 창조하는 것이다.
첫번째 단계는 입구쪽의 벽과 전망용 벽을 따라 유리 플랫폼을 설치함으로써 옥상 공간의 배치를 조정하는 것이다. 이러한 플랫폼은 공간의 크기를 확장시켜줄 뿐만 아니라, 투명한 유리의 사용을 통해 색다른 느낌을 자아내는 동시에 보다 흥미로운 시점과 조망을 제공해준다.

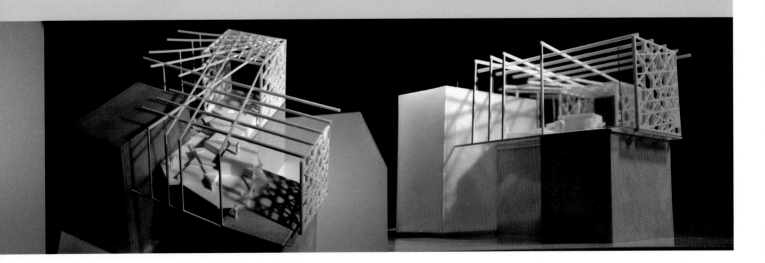

Showing light and airy space

looking at the structure from outside

Cathrine Lampa
cathrinelampa@hotmail.com

The Getaway

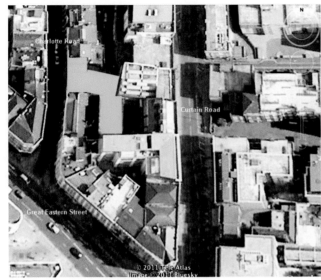

Burbage House is situated on Curtain Road in Hoxton, a young and vibrant area full of art galleries and design companies. In the building there are 5 design studios and the brief was to design the rooftop with the purpose of the people within the building being able to have exhibitions, BBQs, meetings or just to come up to relax or eat their lunch there.

버비지 하우스는 학스턴의 커튼 로드에 위치해 있는데, 젊고 생동감 넘치는 이 지역에는 갤러리와 디자인 회사들이 몰려 있다. 건물에는 5개의 디자인 스튜디오가 입주해 있으며, 우리의 목표는 건물에서 근무하는 사람들이 전시회나 바베큐 파티를 열고, 미팅을 갖거나, 혹은 그냥 올라와서 휴식을 취하고 점심을 먹을 수 있는 옥상을 디자인하는 것이다.

The positions of the neighbouring buildings, there heights and need for light add an ecletic charm. A reflection looking back to the days where intensive industry was the norm in London.

주변 건물들의 높이나 채광권 등을 고려해야 하기 때문에 옥상에서는 절충적인 매력을 느낄 수 있다. 또한 가열찬 산업 발전이 당연시되던 런던의 옛모습을 되돌아볼 수도 있을 것이다.

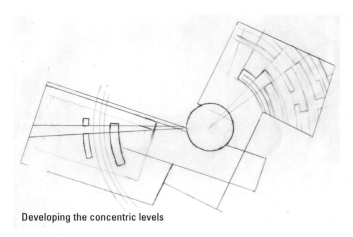

Developing the concentric levels

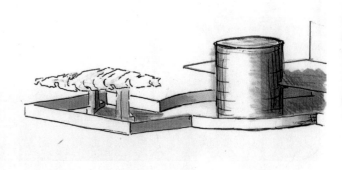

Elevation looking at the levels of roof and tundra.

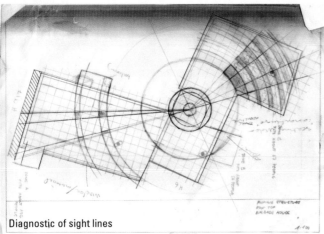

Diagnostic of sight lines

Overview of the concentric lines and hollwed out elements to allow the light to fall within accentuating as rings of light.

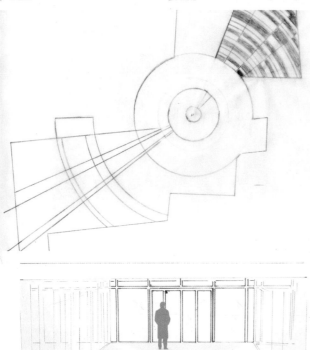

Open views at specific points

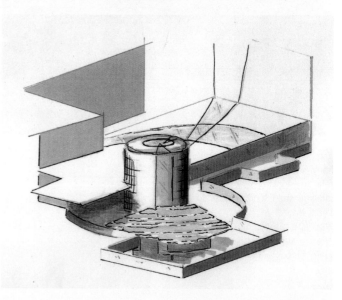

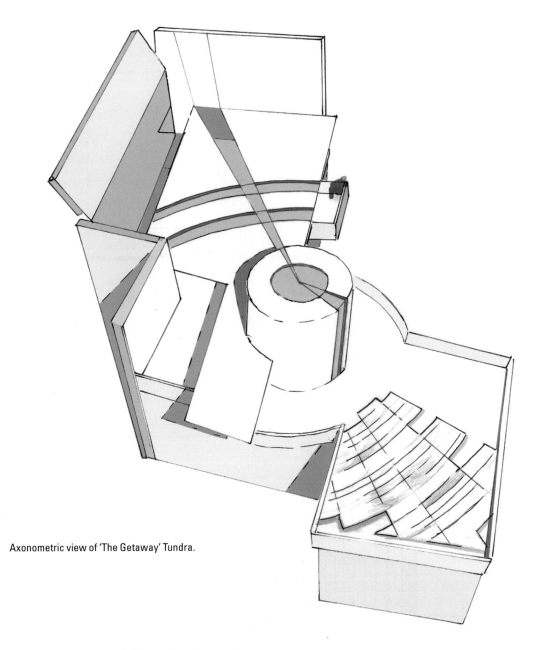

Axonometric view of 'The Getaway' Tundra.

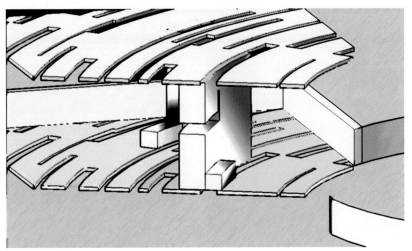

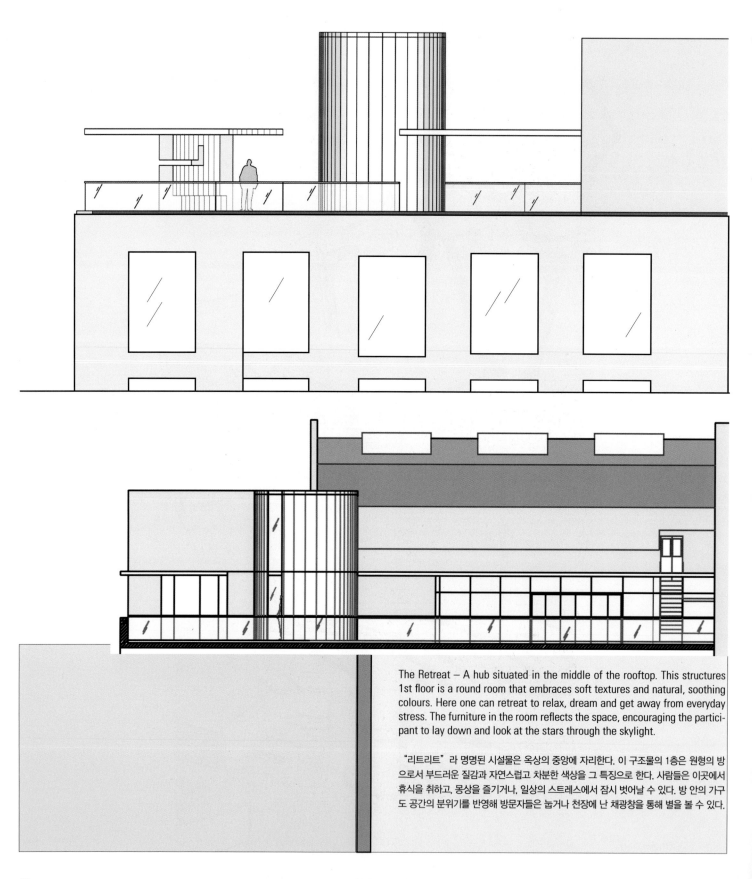

The Retreat – A hub situated in the middle of the rooftop. This structures 1st floor is a round room that embraces soft textures and natural, soothing colours. Here one can retreat to relax, dream and get away from everyday stress. The furniture in the room reflects the space, encouraging the participant to lay down and look at the stars through the skylight.

"리트리트" 라 명명된 시설물은 옥상의 중앙에 자리한다. 이 구조물의 1층은 원형의 방으로서 부드러운 질감과 자연스럽고 차분한 색상을 그 특징으로 한다. 사람들은 이곳에서 휴식을 취하고, 몽상을 즐기거나, 일상의 스트레스에서 잠시 벗어날 수 있다. 방 안의 가구도 공간의 분위기를 반영해 방문자들은 눕거나 천장에 난 채광창을 통해 별을 볼 수 있다.

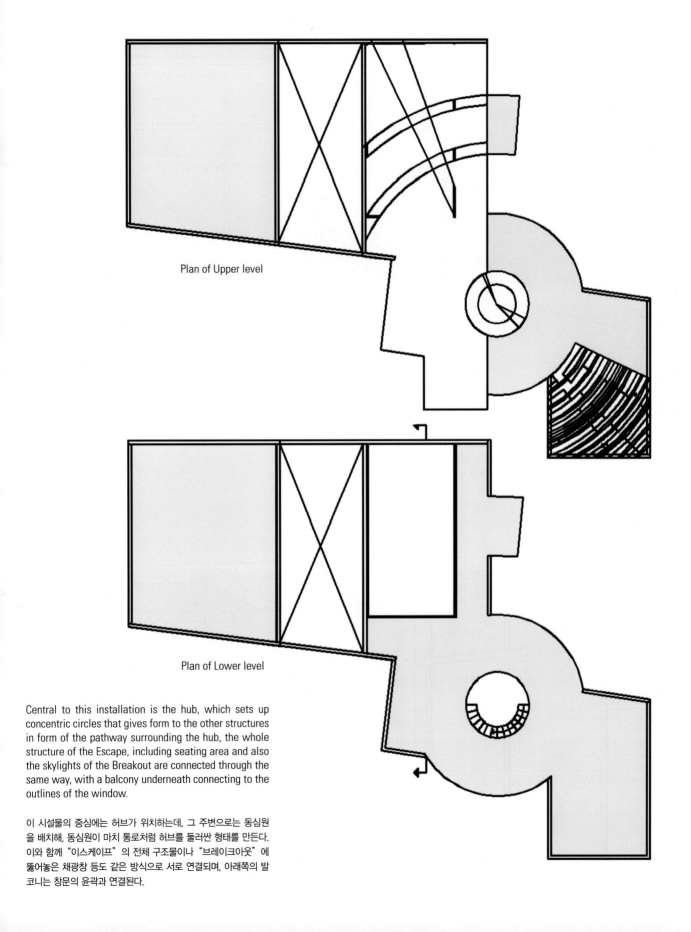

Plan of Upper level

Plan of Lower level

Central to this installation is the hub, which sets up concentric circles that gives form to the other structures in form of the pathway surrounding the hub, the whole structure of the Escape, including seating area and also the skylights of the Breakout are connected through the same way, with a balcony underneath connecting to the outlines of the window.

이 시설물의 중심에는 허브가 위치하는데, 그 주변으로는 동심원을 배치해, 동심원이 마치 통로처럼 허브를 둘러싼 형태를 만든다. 이와 함께 "이스케이프"의 전체 구조물이나 "브레이크아웃"에 뚫어놓은 채광창 등도 같은 방식으로 서로 연결되며, 아래쪽의 발코니는 창문의 윤곽과 연결된다.

The concept is "The Getaway" and the rooftop is designed as 3 unique structures, all providing a new experience of interaction and communication for those utilising the space.
The Breakout – A light and airy exhibition space, where the designers in the building can exhibit their own work, or invite other artists to showcase theirs. This gives an opportunity for designers and artists to meet and make new contacts. All surfaces within this area are flexible, providing   screening or meeting room.

이 컨셉트는 "겟어웨이" 이다. 옥상은 3개의 구조물로 디자인되었으며, 공간을 이용하는 사람들에게 상호작용과 소통이라는 새로운 경험을 가져다 줄 것이다.
경쾌한 전시 공간인 "브레이크아웃" 에서는 건물에 입주해 있는 디자이너들이 자신들의 작품을 전시하거나, 다른 작가들을 초대해 쇼케이스를 열 수 있다.

이를 통해 디자이너들과 아티스트들이 만나 새로운 관계를 맺을 수 있는데, 공간 안의 모든 벽면은 영상물의 상영이나 모임 공간 마련 등을 위해 그 형태를 쉽게 바꿀 수 있다.

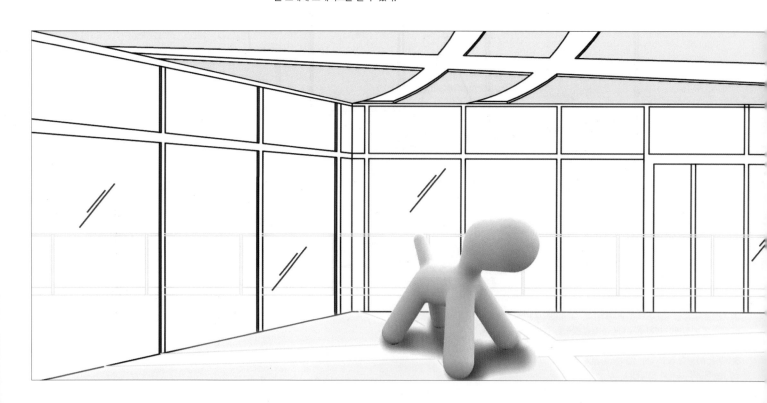

Space allows for the inner exhibits and indeed the people to be placed without restriction being uninhibited by others artistry or architecture as it serves to add character not contrast or or devalue.
The Escape — An airy structure that casts a linear pattern on the floor, camouflaging the person within the space. A Perfect escape!
Open and airey allowing light into the space. Curtain Road and the rooftops of Shoreditch become the backdrop and the character of the gallery.

"이스케이프" 는 바닥에 직선적 패턴이 투사되는 넓직한 공간으로서, 커튼 로드와 쇼어디치의 다른 건물 옥상들이 갤러리의 훌륭한 배경이 된다. 이곳에서는 사람의 모습을 쉽게 구별하기 어려운데, 그야말로 완벽한 일탈이라 할 수 있다.

Cavell Browne
cavellbrowne@yahoo.com
www.cavellbrowne.com

A Woven Frame Narrative

Positioned in the heart of Shoreditch triangle is Burbage House, a twentieth century building situated on the west side of Curtain road: Named in tribute to the family inherently linked with the theatrical world of Shakespeare's day. The 'Theatre' was London's first purpose built playhouse (built by James Burbage in 1576) it lasted for approximately 21 years. One year later 1577 a second playhouse, The Curtain, was built in Shoreditch hence the name Curtain Road.

쇼어디치의 중심부에 위치한 버비지 하우스는 커튼 로드의 서쪽에 자리잡은 20세기에 세워진 건물이다. 그 이름은 셰익스피어시대의 연극 활동과 연관이 깊은 한 집안을 기리기 위해 붙여진 것이다.
"시어터" 는 1576년 제임스 버비지에 의해 세워진 런던 최초의 극장으로서 약 21년간 명맥을 유지했다. 1년 뒤인 1577년 두번째 극장인 "커튼" 이 쇼어디치에 설립되는데, 커튼 로드란 명칭은 이 극장에서 따온 것이다.

Burbage House Curtain Road

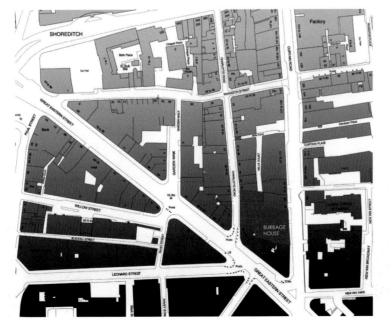

'Should the buildings we live/work structurally reflect the way we visually communicate who and what we are, in the hope of forging connections within the communities that we live, especially in this ever increasingly critical day?'

"우리가 일하는, 살고있는 건물들이, 커뮤니티 내부에 소통을 이루어낸다는 희망과 함께, 우리가 어떻게 의사소통을 하는지, 우리가 누구인지를 구조적으로 반영할 수 있을까 ? 그것도 나날이 사람들이 냉담해져 가는 이런 시절에 ?"

Through the analysis of the companies situated within Burbage House a structural weave configuration has been formed; translating this weave pattern from its geometrical form, in opening it up from its formation, a framework for architectonic perspective has come into view.
The potentiality of fusing architecture Architectural design is becoming more individualistic and adaptive to our requirements, desires and pleasures. with modern woven craft techniques realises a new and personal visual dialogue.

버비지 하우스에 입주한 업체들에 대한 분석을 통해 구조에 있어서의 무늬 구성을 만들어냈다. 기하학적 형태로부터 이러한 무늬 패턴을 해석해냄으로써, 건축학적 시각의 기본 틀을 수립할 수 있었다.

The woven narrative was inspired by the dialogue created by the companies situated within Burbage House, in relation to its historical surroundings; translating this installation from a structural standpoint into a frame narrative.

우븐 내러티브(woven narrative)라는 개념은 버비지 하우스에 입주한 기업들이 나누는 대화에서 영감을 얻어 이 지역의 역사적 배경을 고려해 만들어낸 것이다. 구조적인 관점에서 이 건물을 해석해냄으로써 전체적인 틀로서의 프레임 네러티브를 만들어낼 수 있었다.

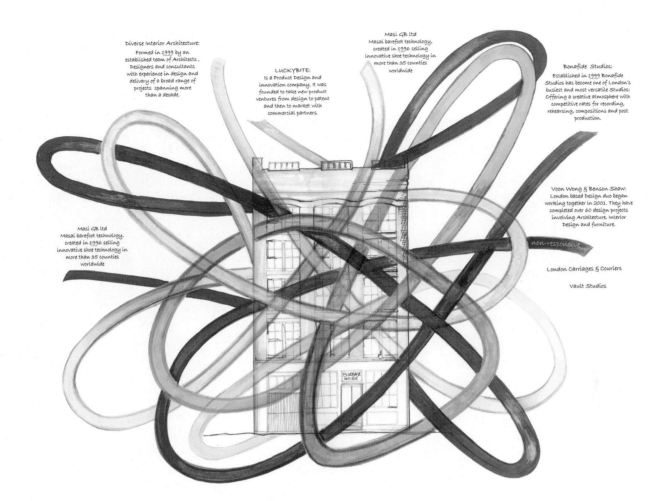

Diverse Interior Architecture:
Formed in 1999 by an established team of Architects, Designers and consultants with experience in design and delivery of a broad range of projects spanning more than a decade.

LUCKYBITE:
Is a Product Design and innovation company. It was founded to take new product ventures from design to patent and then to market with commercial partners.

Masi GB ltd
Masai barefoot technology, created in 1996 selling innovative shoe technology in more than 35 counties worldwide

Bonafide Studios:
Established in 1999 Bonafide Studios has become one of London's busiest and most versatile Studios: Offering a creative atmosphere with competitive rates for recording, rehearsing, compositions and post production.

Voon Wong & Benson Shaw:
London based Design duo began working together in 2001. They have completed over 60 design projects involving Architecture, Interior Design and furniture.

non-responsive

London Carriages & Couriers

Vault Studios

Masi GB ltd
Masai barefoot technology, created in 1996 selling innovative shoe technology in more than 35 counties worldwide

Etymology of Shoreditch

According to an ancient legend ballad, Elizabeth Shore, known as Jane Shore who was the most noted of all of king Edward IV mistresses, died in a ditch! A more likely argument is that 'Soersditch' (sewer ditch) was an ancient drain or watercourse in what was possibly a boggy area.

쇼어디치라는 지명의 유래

전설적인 고대 발라드에 의하면, 에드워드 4세 국왕의 여인들 중 가장 널리 알려진 제인 쇼어, 일명 엘리자베스 쇼어는 배수로에서 죽었다고 한다. "쇼어스디치" 라는 말이 고대에 배수로 따위를 지칭했다는 주장은 보다 설득력이 있다.

Demography

In 1801, the civil parishes that form the modern borough had a total population of 14,609. This rose steadily throughout the 19th century as the district became built up, reaching 95,000 in the middle of the century. With the arrival of the railway, the rate of population increased reaching nearly 374,000 by the turn of the century.

The London Borough of Hackney was formed in 1965 from the area of the former metropolitan boroughs of Hackney, Shoreditch and Stoke Newington. The 2001 census gives Hackney a population of 202,824. The population is ethnically diverse. Of the resident population, 89,490 (40%) people describe themselves as White British. 30,978 (15%) are in other White ethnic groups, 63,009 (30%) are Black or Black British, 20,000 (9.4%) are Asian or Asian British, 8,501 (4%) describe themselves as 'Mixed', and 6,432 (3%) as Chinese or Other.

During the 17th century, Shoreditch saw an influx of immigrants' refugees the Huguenot (French Protestants). Fleeing from the religious persecution in their homeland they added a new dimension to the culture and history of the area. The Huguenot established a major weaving industry in and around the surrounding areas of Shoreditch and Spitafield. During the 19th century Shoreditch also became the centre of the furniture industry; many pieces are today exhibited in the Geffrye Museum on Kingsland Road. Towards the end of 19th Century saw the decline of both the textile and furniture industries, with Shoreditch fast becoming renowned for crime, prostitution and poverty

인구통계

1801년 오늘날의 자치구를 이룰 법한 이 지역의 총인구는 14609명이었다. 19세기 내내 인구는 점진적으로 증가하는데, 19세기 중엽에 이르자 95,000명에 이르게 된다. 철도의 개설과 더불어 인구는 급격하게 증가하는데, 20세기에 접어들 무렵 거의 374,000명에 이르게 된다.

1965년 해크니, 쇼어디치, 스토크 뉴잉턴의 자치구를 통합해 런던의 해크니 자치구가 만들어진다. 2001년 인구총조사는 해크니 지역의 총인구를 202,824명으로 추산하고 있으며, 이 지역의 인종적 다양성을 보여주고 있다. 주민들 가운데 89,490명(40%)은 자신들을 백인 영국인이라 표현했고, 30978명(15%)는 다른 인종의 백인이라 답했다. 63,009명(30%)은 자신들을 흑인 또는 흑인 영국인이라 답했으며, 20,000명(9.4%)은 아시아인 또는 아시아계 영국인이라 답했다. 8,501명(4%)은 자신들을 "혼혈" 이라 칭했고, 중국인 또는 기타 인종이라 응답한 사람들은 6,432명(3%)이었다.

17세기에 프랑스의 신교도인 위그노들이 쇼어디치에 몰려오게 된다. 고국에서의 종교적 박해를 피해 도망쳐온 이들은 이 지역의 역사와 문화에 새로운 차원을 더하게 된다. 위그노들은 쇼어디치 및 스피탈필드, 그리고 그 인근 지역에 주요 방직 공장을 세웠다.

19세기에 쇼어디치는 가구 산업의 중심지가 되기도 한다. 이 때 만들어진 많은 가구들이 킹스랜드 로드의 제프리 박물관에 전시되고 있다. 19세기 말엽으로 접어들면서 쇼어디치 지역의 섬유 및 가구 산업은 모두 쇠퇴의 길을 걷게 되고, 이 무렵부터 이 지역은 범죄, 매춘, 그리고 빈곤으로 널리 알려지게 된다.

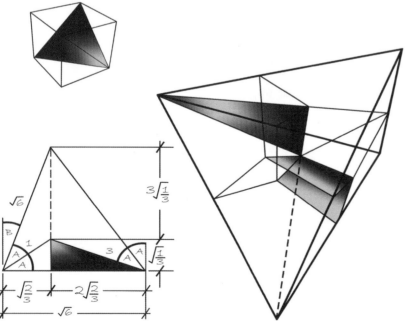

This project is a design exploration into how the qualitative data of the companies situated within Burbage House, has taken form and woven an architectural dialogue. Using the decorative macrogauze woven technique developed by Peter Collingwood, the installation breaks down the 'traditional' walls of a building into a spatial frame story.

From the inside out the shapes build upon one another; both materially and geometrically, blurring the boundary between the interior and the exterior thus creating, through the specificity of material used a series of distinctive descending spaces.

이 프로젝트는 디자인 탐구로서 버비지 하우스에 입주한 업체들에 대한 질적 데이터가 어떻게 형태를 갖게 되는지, 그리고 어떻게 건축적 대화를 엮어내는지에 대해 연구할 것이다. 피터 콜링우드가 개발한 장식적 매크로가우즈 무늬 기법을 이용해, 설치물은 "전통적"인 건물의 벽을 허물어 공간적 이야기 구조를 만들어낸다.

안쪽에서부터 바깥쪽으로 형체들이 질료상으로, 그리고 기하학적으로 차곡차곡 쌓여간다. 이를 통해 안과 밖의 경계는 불분명해진다.

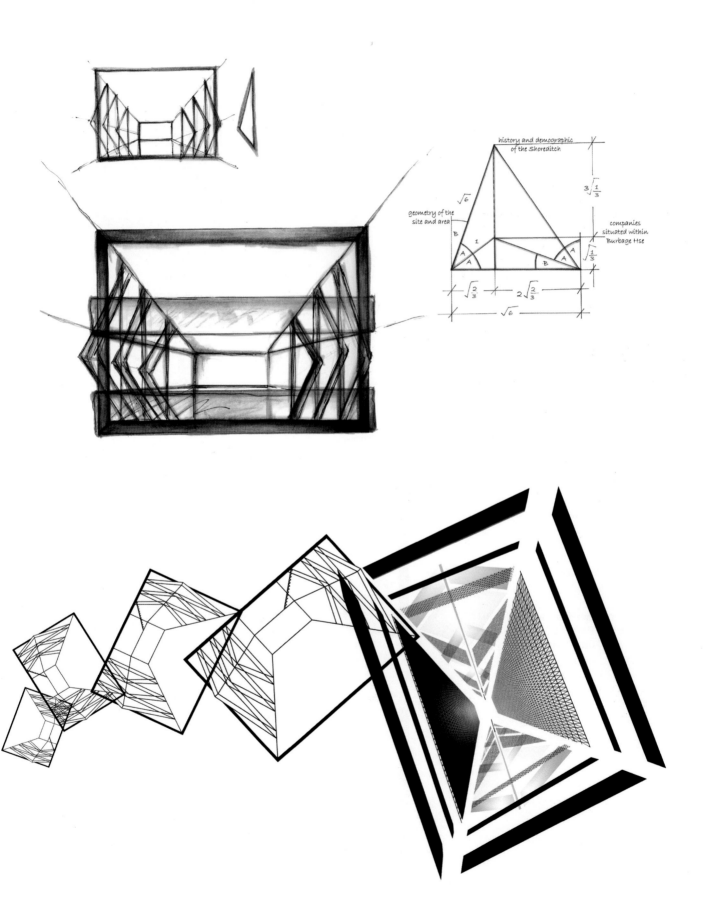

history and demographic
of the Shoreditch

geometry of the
site and area

companies
situated within
Burbage Hse

$\sqrt{6}$

$3\sqrt{\frac{1}{3}}$

$\sqrt{\frac{1}{3}}$

B

1

A
A

B

A

$\sqrt{\frac{2}{3}}$

$2\sqrt{\frac{2}{3}}$

$\sqrt{6}$

Predominantly a working class area, Shoreditch has, in recent years, been gentrified by the creative industries and those who work in them. Located on New Inn Street is the London College of Fashion (LCF); which was created in 1967 from the merger of the Shoreditch Institute Girls Trade School (founded 1906) and the Trade School for Girls, Barrett Street (founded 1915) forming a single college for the garment trade, subsequently now there is an active trade in fashion design/studios. A sculpture erected on Bethnal GreenRoad, E2, near the junction with Sclater Street aptly named Threads ; made by a brother and sister who used a century-old rope to make it, is intended as a visual reminder of the connection the area has with the garment and textile industries. Former industrial buildings have been converted to offices and flats, while Curtain Road and Old Street are notable for their clubs, pubs, bars, restaurants and media/print businesses which offer a variety of venues to rival those of the West End.

주로 노동자 계층이 거주하는 쇼어디치는 최근 몇 년새 창조적 기업체 및 그곳에 종사하는 사람들에 힘입어 그 수준을 끌어올릴 수 있었다. 뉴인 스트리트에는 런던 패션 컬리지(LCF)이 자리잡고 있는데, 이 학교는 1967년 쇼어디치 여자 직업학교(1906년 설립)와 배럿 스트리트에 위치한 여자 직업학교(1915년 설립)의 합병을 통해 탄생하였다. 베스날 그린 로드에 세워진 동상의 이름이 "스레즈(옷감 짜는 실)"인 것은 매우 적절하다 할 수 있는데, 이를 통해 이 지역이 섬유 및 의복 산업과 밀접한 관련을 맺고있다는 사실을 시각적으로 확인할 수 있다.

이전의 공장건물들은 사무실이나 아파트로 개조되었고, 커튼 로드와 올드 스트리트에는 클럽, 펍, 바, 레스토랑, 그리고 미디어 및 인쇄 업체들이 들어서 웨스트 엔드와 견줄만한 다양성을 확보하게 되었다.

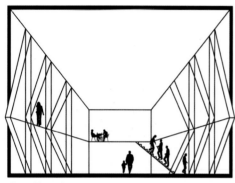

Front Elevation

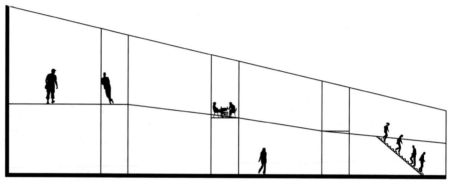

Side Elevation

Juxtaposition of the old against new...

The proposal of Textiles and Fibre to be used in the installation on the rooftop of Burbage House reflects the rich history that Shoreditch has with the textile Industry. With the emergence of new technology, textile and fibre can no longer be dismissed as eccentric, marginal or idiosyncratic pastime. In fact Textile Artists have been crossing all media boundaries with many of their work reflecting a historic, cultural and a personal dialogue.

신구(新舊)의 병치

버비지 하우스의 옥상 구조물에 직물과 섬유를 사용하고자 하는 것은 쇼어디치와 섬유산업 간의 오랜 역사적 관계를 반영하기 위해서이다. 새로운 테크놀러지의 등장 덕분에, 직물과 섬유는 더 이상 미미하거나 색다른 소재로 정도로 취급되지 않게 되었다. 사실상 직물 아티스트들은 모든 매체의 경계를 뛰어넘으면서 역사적, 문화적, 개인적 이야기를 담은 다양한 작품들을 선보이고 있다.

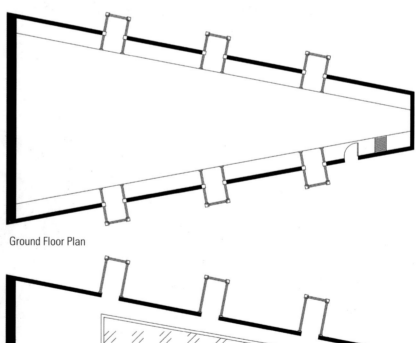

Ground Floor Plan

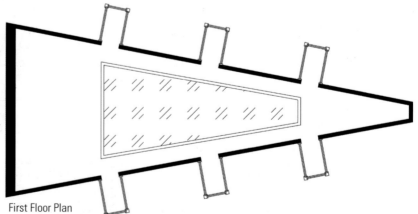

First Floor Plan

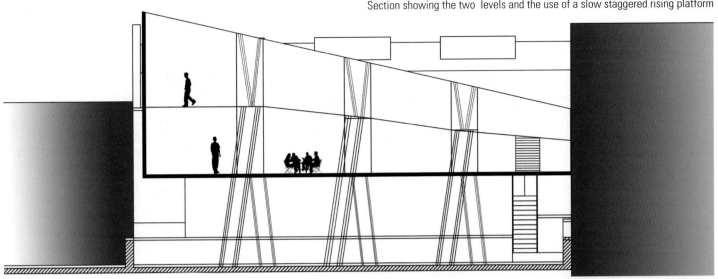

The installation encompasses the techniques and craftsmanship of weaving which acknowledges the evolution of the woven fabric. Research found that French weavers in the 15th century had a repertory of tales; (french tale-teller trouvere), assembled a collection of stories entitled Les Évangiles des Quenouilles Spinners Tales known as a 'frame story'. A frame story is a narrative technique which leads readers from the first story into the smaller stories within it. Therefore the installation, with modern technologies to achieve a blending of the new and advance traditions, generates an emotional response that is connected to something we know and recognise but has been adapted in an innovative way.

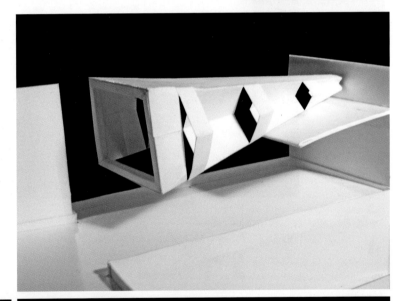

구조물은 직조와 관련된 모든 기술과 재주를 아우른다. 연구에 따르면 15세기 프랑스의 직공들은 여러 개의 이야기들로 구성된 이야기 레퍼토리를 갖고 있었다고 한다. 이러한 레퍼토리는 액자식 구성으로도 알려져있는데, 액자식 구성은 일종의 서사 기법으로서 독자들은 첫번째 이야기가 시작되면 그 안에 들어있는 다른 작은 이야기들로 차차 움직여가게 된다. 이와 마찬가지로 신구의 조화를 위해 현대적 기술이 사용된 본 구조물을 통해, 이미 잘 알고 있지만 혁신적 방법을 통해 변용된 것에 대해 우리가 갖게 되는 감정적 반응을 자아낼 수 있는 것이다.

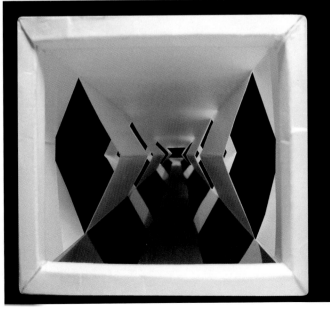

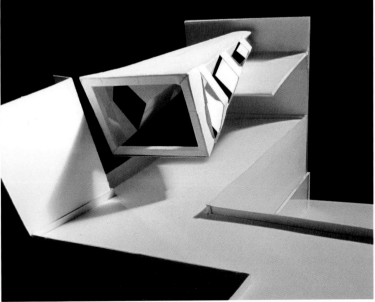

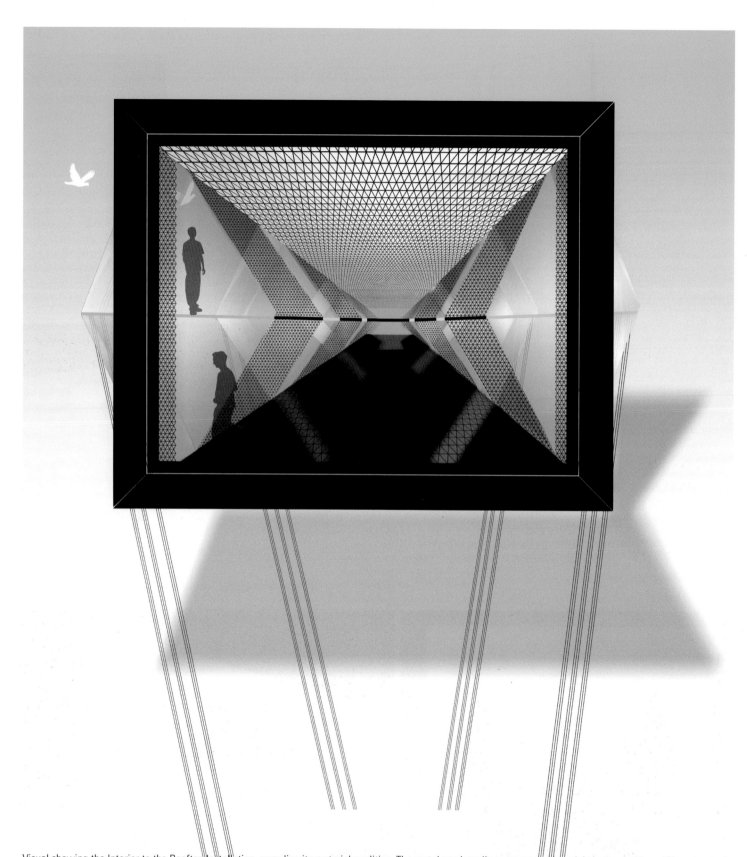

Visual showing the Interior to the Rooftop Installation, revealing its material qualities. The metal mesh walls compress and straighten to provide a different experience for differing functions which relate to time and light values.

철망으로 만든 벽은 수축하고 또 팽창하기도 하면서, 시간과 빛에 따라 변화하는 각기 다른 기능에 대한 각기 다른 경험을 제공한다.

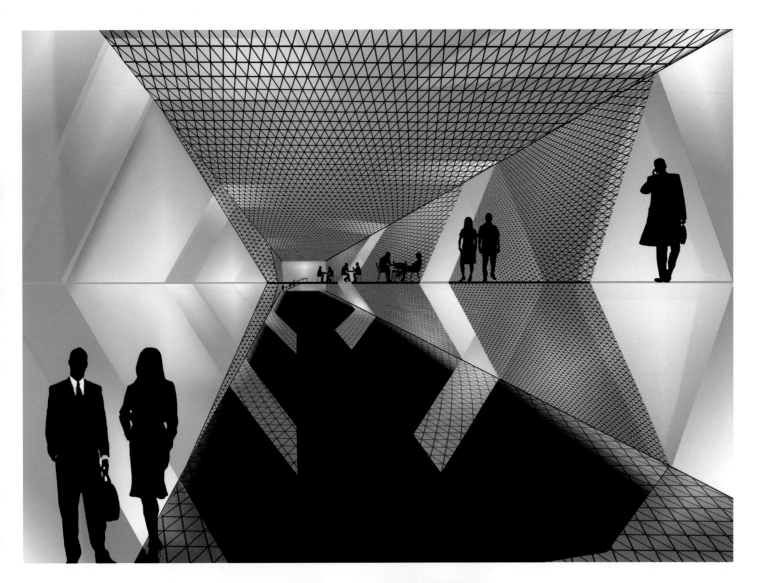

Rooftop Subterraneanism

A New Perspective on Spatial Subterranean: By elevating the installation above the rooftop, the rooftop space has been affected but not compromised, offering maximum flexibility to the inhabitants. The threads of the woven structure once made contact with the floor of the rooftop, morphs into shafts of solar lights translating the woven dialogue into a visual statement where images/codes can be embedded, allowing the rooftop to be used in several option i.e. outdoor conferencing/exhibitions or barbecues.

옥상의 지하화

새로운 구조물을 옥상 위에 설치하게 되면 어떤 식으로든 옥상 공간에 영향을 주게 되지만, 이용자들에게 최대한의 가변성을 제공함으로써 본래의 기능이 훼손되는 것을 피할 수 있다.

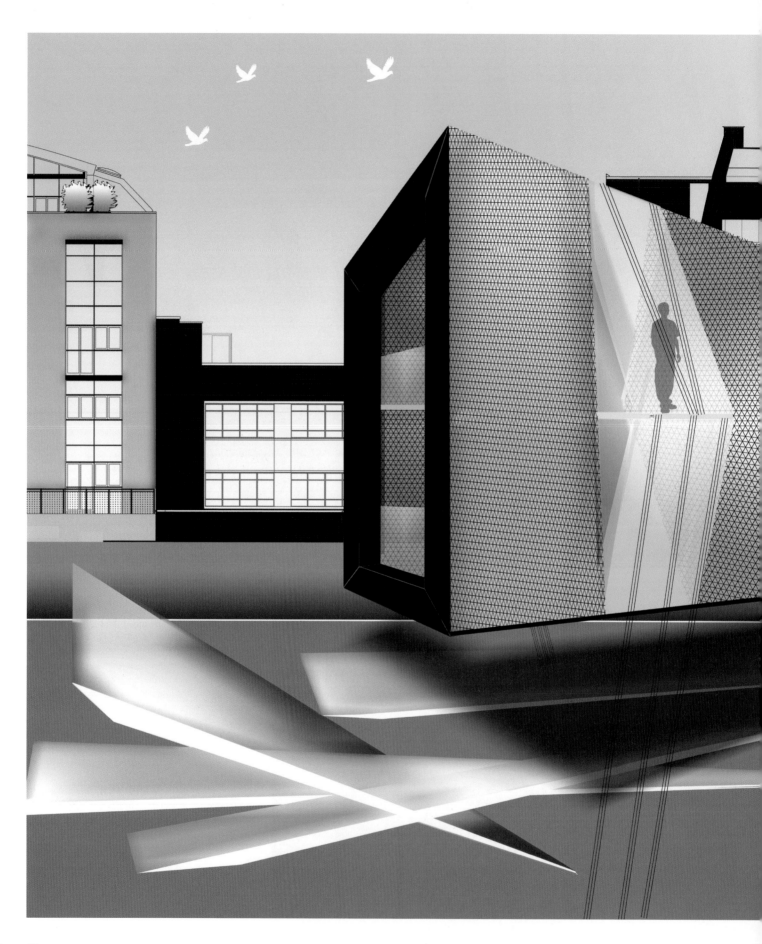

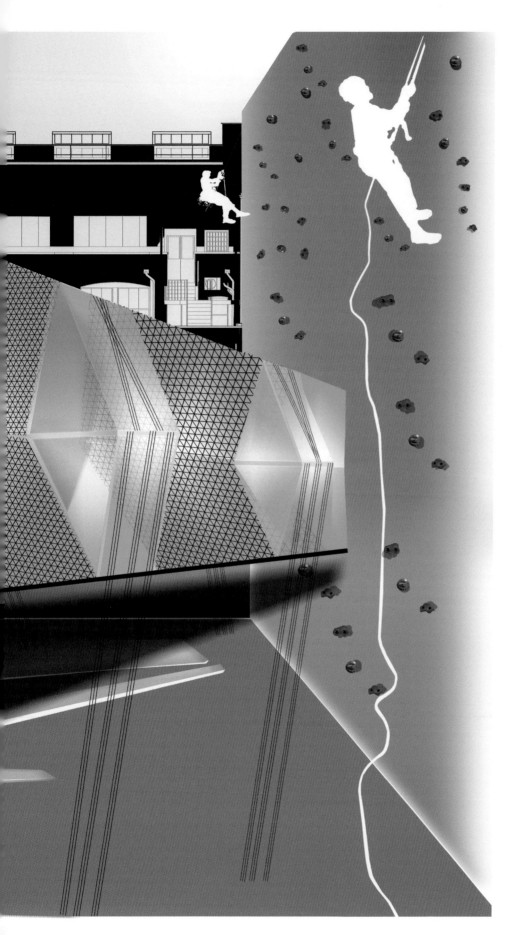

The building's most prominent feature is the energy woven wirecloth membrane which constitutes the outermost layer. Creating not only a diffused lighting and a comfortable climatized zone inside but also, through its folding and sometimes reflective, sometimes translucent surface, accentuates the diamond cut appearance of the structure.

건물에서 가장 도드라진 부분은 건물의 가장 바깥쪽 층을 구성하고 있는 촘촘하게 짜인 에너지 그물막이라 할 수 있다. 이는 확산 조명과 내부의 편안한 공간을 만들어낼 뿐만 아니라, 때로는 빛을 반사하고, 때로는 반투명한 표면을 드러냄으로써 구조물의 다이아몬드 커트 외관을 더욱 돋보이게 한다.

Perspective Looking to Charlotte Road
In the evening the entire structure Lights up providing both ambient light within and a stunning architectural effect. The neighbouring wall is used as a climbing wall.

샬롯 로드 방향
저녁 무렵이 되면 구조물 전체에 불이 들어오면서 내부에는 은은한 조명이 비추고, 외부에는 놀라운 건축적 효과가 나타나게 된다. 이웃한 건물벽은 암벽등반용 인공 암벽으로 이용된다.

Charis Christodoulou
ch.charis@yahoo.com

ATARAXIA

CURTAIN ROAD

Shoreditch Area
Some roads lack identity, forgotten by the city and the fashionable. However the newly formed businesses benefit from low rents from this lack of attention.

BURBAGE HOUSE

Entrance to Burbage House 4 Storey Industrial building.

ROOFTOP
facing at Charlotte Road

Flat Roofs- Typical of warehouses of this type - opposite are pitched roofs belonging to a slightly older architecture in Charlotte Road.

ROOFTOP
facing towards Curtain Road

Looking towards a more contemporary development-stark contrast to the typical buildings of this area.

ROOFTOP
facing the middle platform

The Rooftop is totally surrounded by safety walls-building houses another escape stair.

Sketch of centre rooftop section looking to Charlotte Road

Sketch of first rooftop section over looking Burbage House and the Stairwell void to the right

Bridge crossing over rooftop platforms

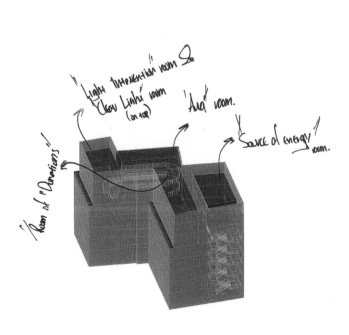

Boxed model showing diagnostic placement of ideas

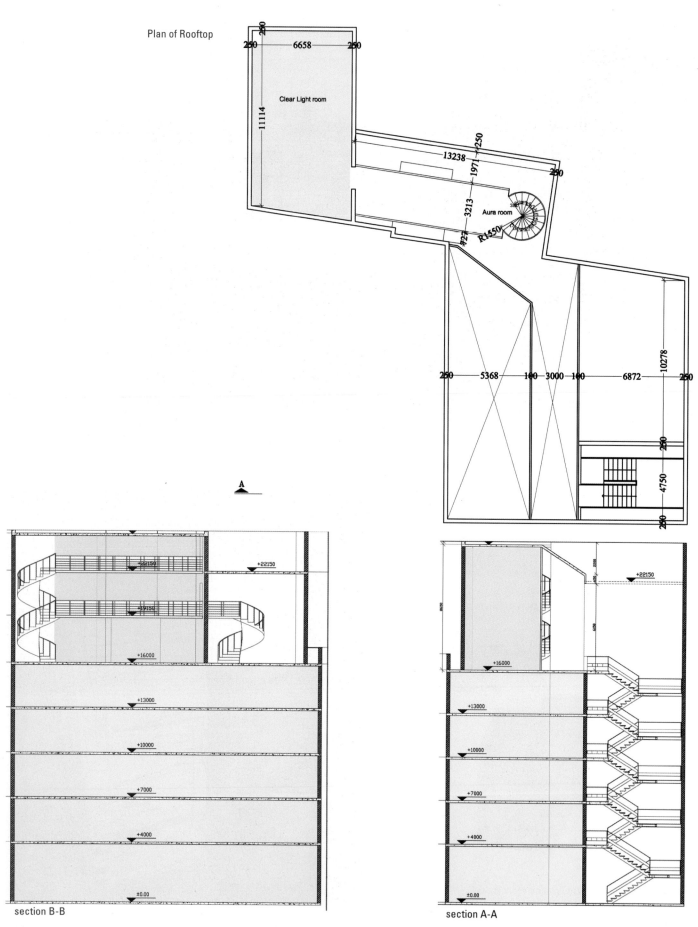

Plan of Rooftop

Clear Light room

Aura room

section B-B

section A-A

Tinted Shadows

Tinted patternationation

region C-
light intervention
space
natural light produces
shadows from a pattern of concrete
pieces attached to glass walls
& clear light room (on top)
open air space with purest
form of light, as there is no
light manipulation

Roofplan 1:200
forms of light manipulation

aura based colours are exposed
in the form of reflective glass
panels, three angles are available to the visitor: ground floor,
a higher 3-meter platform
(leads to light intervention space)
and
a second platform 6-meters
(leads to the clear light space)

region A-
room of donations
artificial light produces
shadows from hanging in-
stallation where visitors
donate their objects

aura corridor
region B-

region A-
initially a gap
source of
energy space
dark room brightened
by illuminative white
pipes

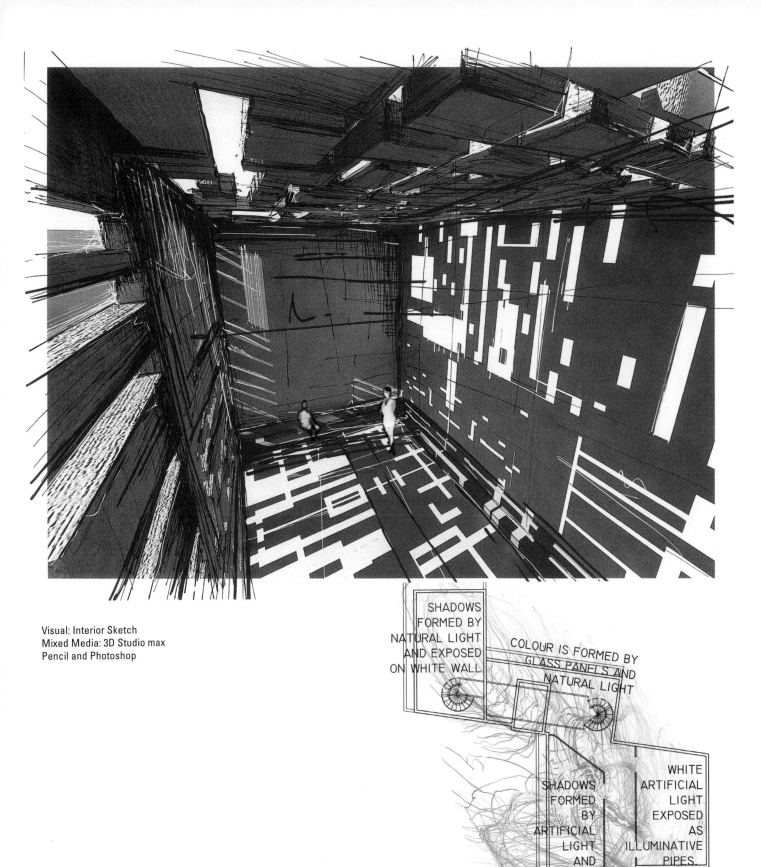

Visual: Interior Sketch
Mixed Media: 3D Studio max
Pencil and Photoshop

SHADOWS FORMED BY NATURAL LIGHT AND EXPOSED ON WHITE WALL

COLOUR IS FORMED BY GLASS PANELS AND NATURAL LIGHT

SHADOWS FORMED BY ARTIFICIAL LIGHT AND VARIETY OF OBJECTS

WHITE ARTIFICIAL LIGHT EXPOSED AS ILLUMINATIVE PIPES.

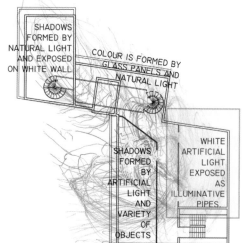

SHADOWS
FORMED BY
NATURAL LIGHT
AND EXPOSED
ON WHITE WALL

COLOUR IS FORMED BY
GLASS PANELS AND
NATURAL LIGHT

SHADOWS
FORMED
BY
ARTIFICIAL
LIGHT
AND
VARIETY
OF
OBJECTS

WHITE
ARTIFICIAL
LIGHT
EXPOSED
AS
ILLUMINATIVE
PIPES.

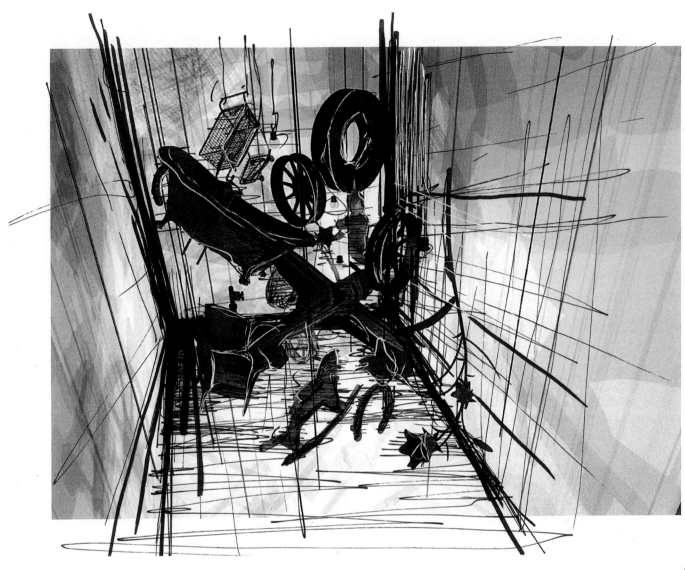

Room of donations: artificial light produces shadows from hanging installation.

where visitors donate their objects.

Source of energy space: dark room brightened by illuminative white pipes.

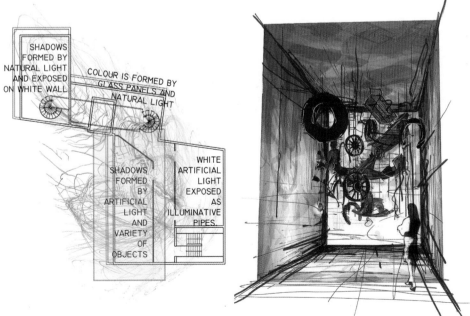

SHADOWS FORMED BY NATURAL LIGHT AND EXPOSED ON WHITE WALL

COLOUR IS FORMED BY GLASS PANELS AND NATURAL LIGHT

SHADOWS FORMED BY ARTIFICIAL LIGHT AND VARIETY OF OBJECTS

WHITE ARTIFICIAL LIGHT EXPOSED AS ILLUMINATIVE PIPES.

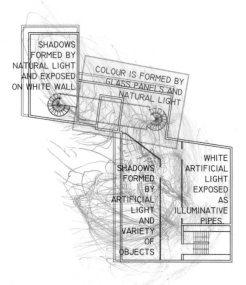

SHADOWS FORMED BY NATURAL LIGHT AND EXPOSED ON WHITE WALL

COLOUR IS FORMED BY GLASS PANELS AND NATURAL LIGHT

SHADOWS FORMED BY ARTIFICIAL LIGHT AND VARIETY OF OBJECTS

WHITE ARTIFICIAL LIGHT EXPOSED AS ILLUMINATIVE PIPES.

Located within the shoreditch area, Ataraxia is an office extension retreat area which aims to treat your eyes serenely.
The space is an approach to regain your initial tranquillity, vanished during work. exposing a sentiment of belonging and mental clarity, space was manipulated to create a meaningful & respectful space both for the visitor and building.the journey is based on keeping an honest approach to the design.

쇼어디치 지역에 자리잡고 있는 아타락시아는 사무실의 연장선상에 놓인 휴식 공간으로서 당신의 시각을 평화롭게 어루만지는 것을 목표로 한다.
이 공간을 통해 당신은 업무에서 잠시 벗어나 마음의 평안을 얻고, 영혼의 정화를 경험하게 될 것이다. 방문자와 입주자 모두에게 의미있고 경의를 표할만한 공간을 만들기 위해 공간에 많은 변화를 가하였고, 디자인에 대해 정직한 태도로 접근한다는 것을 모토로 하였다.

The visitor should be transported to an ecstatic atmospheric space empha-
sising a mixture of raw based materials and exposing light & shadow interference.
The space is manipulated by light and takes a variety of forms as each individual's state of calmness can be realised differently from someone else's.

방문자들은 가공되지 않은 소재와 빛과 그림자를 근간으로 하는 황홀한 공간을 결험하
게 된다.
공간은 빛에 의해 변화하면서 다양한 모습을 띄게 되는데, 사람들 마다 마음의 평정을 얻
는 방식이 다르기 때문이다.

Aura corridor: aura based colours are exposed in the form of reflective glass
Panels. Three angles are available to the visitor: ground floor,
A higher 3-meter platform (leads to light intervention space)
And a second platform 6-meters (leads to the clear light space)

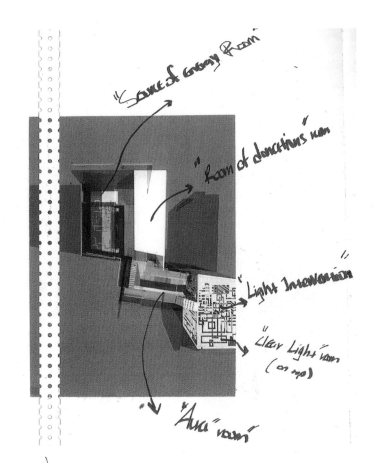

Gbubemi Eyetsemitan
gbubemee@hotmail.com

Curtain Call

About Burbage House
Burbage House is the site of the proposed roof installation which is located on Curtain road in Shoreditch, London. The area is presently predominantly a hotspot for design orientated companies

The Concept
The concept is as a result of research on the Burbage family and their exploits in the world of British theatre. James Burbage being the father of two sons Richard and Cuthbert Burbage is the pioneer of theatre in London and the owner of London's first theatre houses Blackfriars Theatre and The Theatre which was eventually dismantled and its parts relocated to build the Globe Theatre.

The People
This building itself is presently occupied by a variety of differently disciplined design businesses, from film to interiors and architecture

버비지 하우스에 대하여
버비지 하우스는 옥상 구조물 설치를 의뢰받은 장소로서 런던 쇼어디치 지역 커튼 로드에 위치하고 있다. 이 지역에는 현재 디자인 관련 회사들이 주로 몰려있다.

컨셉트
디자인의 컨셉트는 버비지 가문과 영국 연극계에서 이들이 이룬 업적에 대한 연구를 바탕으로 구상하였다. 리차드와 커스버트, 두 아들의 아버지였던 제임스 버비지는 런던 연극계의 선구자였으며 런던 최초의 극장인 블랙프라이어스 시어터와 더 시어터의 소유주이기도 했다. 후자의 경우 이후 철거되어 글로브 시어터를 건설하는 자재로 사용되기도 한다.

GLASS TUBE

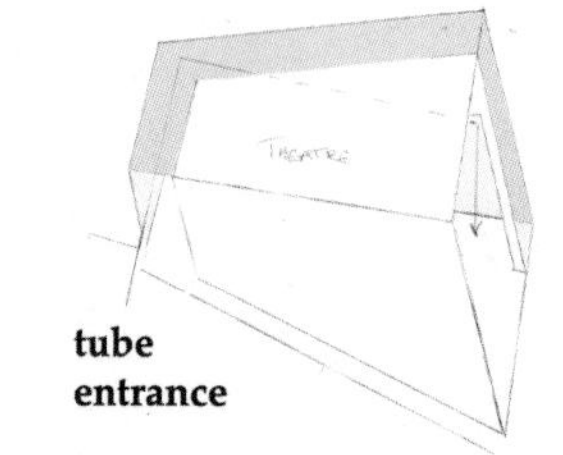

tube entrance

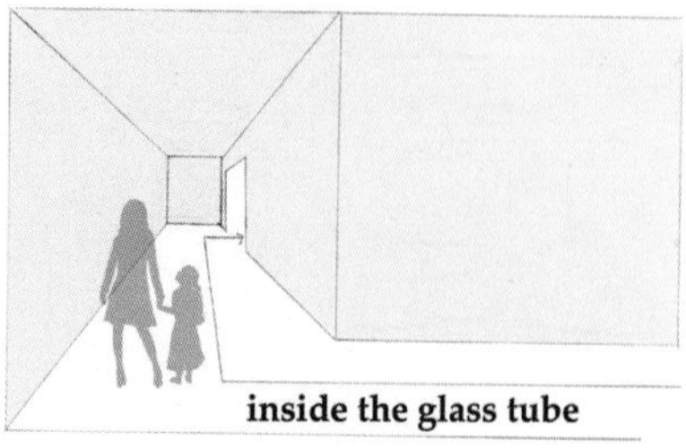

inside the glass tube

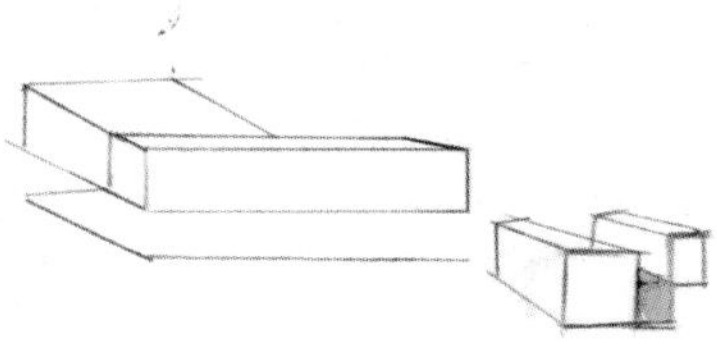

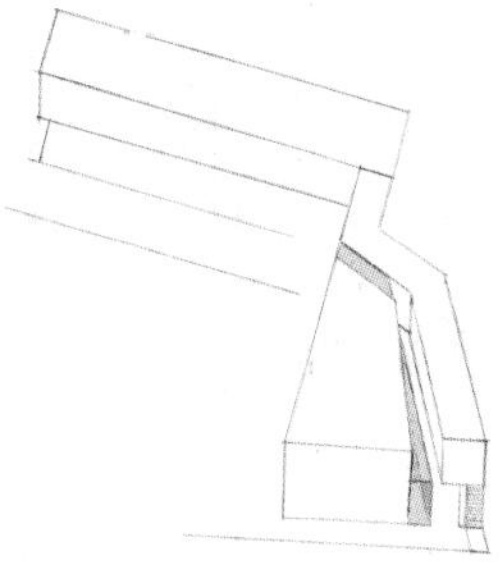

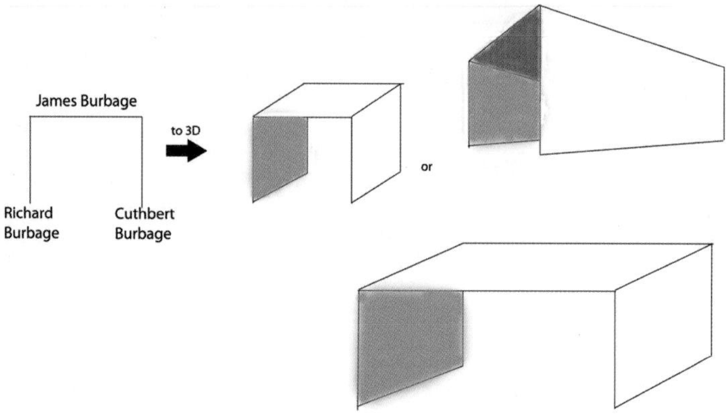

James Burbage

→ to 3D →

Richard Burbage   Cuthbert Burbage

or

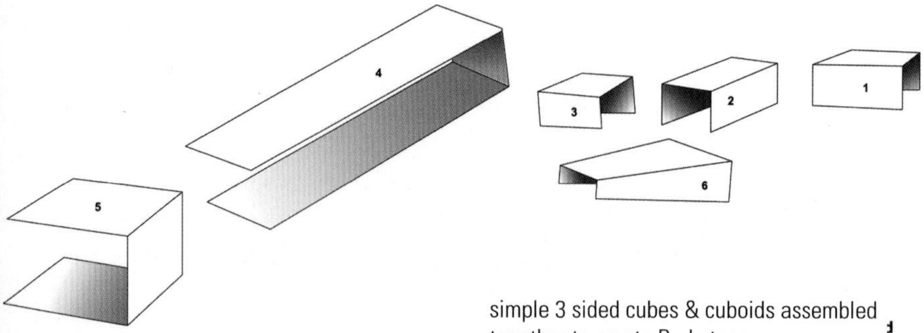

4

3   2   1

5

6

simple 3 sided cubes & cuboids assembled together to create Burbstage.

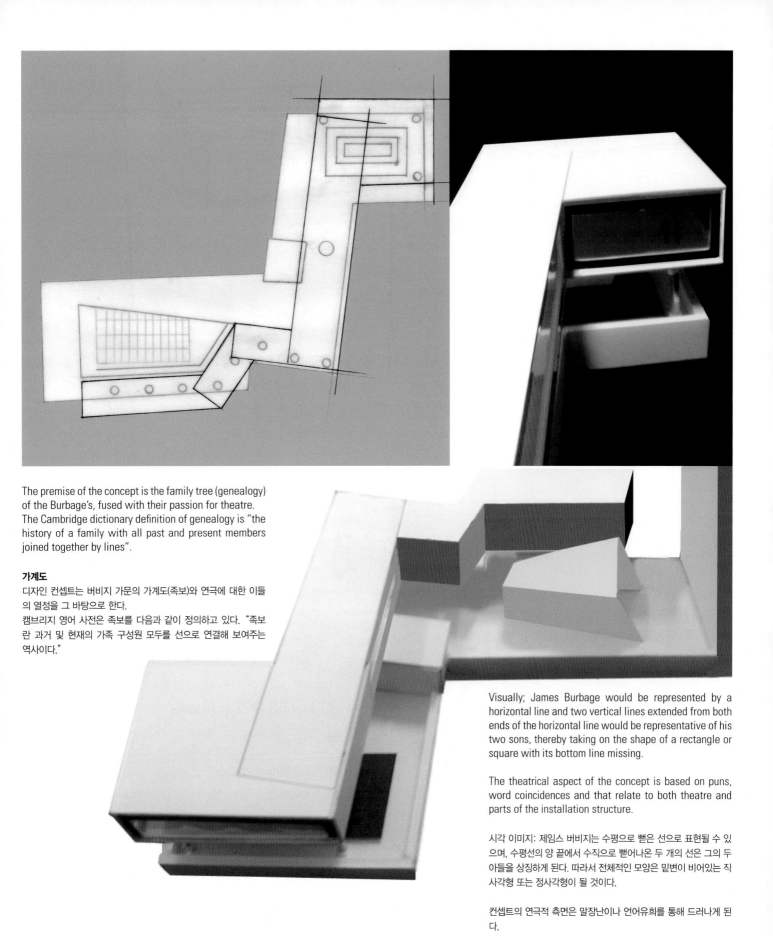

The premise of the concept is the family tree (genealogy) of the Burbage's, fused with their passion for theatre. The Cambridge dictionary definition of genealogy is "the history of a family with all past and present members joined together by lines".

가계도

디자인 컨셉트는 버비지 가문의 가계도(족보)와 연극에 대한 이들의 열정을 그 바탕으로 한다.
캠브리지 영어 사전은 족보를 다음과 같이 정의하고 있다. "족보란 과거 및 현재의 가족 구성원 모두를 선으로 연결해 보여주는 역사이다."

Visually; James Burbage would be represented by a horizontal line and two vertical lines extended from both ends of the horizontal line would be representative of his two sons, thereby taking on the shape of a rectangle or square with its bottom line missing.

The theatrical aspect of the concept is based on puns, word coincidences and that relate to both theatre and parts of the installation structure.

시각 이미지: 제임스 버비지는 수평으로 뻗은 선으로 표현될 수 있으며, 수평선의 양 끝에서 수직으로 뻗어나온 두 개의 선은 그의 두 아들을 상징하게 된다. 따라서 전체적인 모양은 밑변이 비어있는 직사각형 또는 정사각형이 될 것이다.

컨셉트의 연극적 측면은 말장난이나 언어유희를 통해 드러나게 된다.

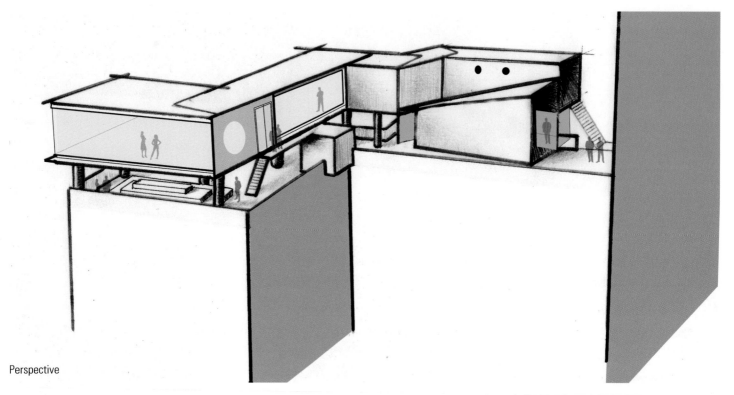

Perspective

The transformation of these three lines or incomplete square or rectangle into a three dimensional image becomes a cube or cuboid with three of its sides missing, which is the predominant shape in the installation. The arrangement of these simple shapes on the site is achieved by mounting, aligning and joining one piece to another in a spatially aware manner to make the space accessible to more people with the projection screen as the central focus.

앞서 설명한 3개의 선 또는 불완전한 사각형을 3차원으로 변환시키면 3개의 면이 비어있는 육면체를 얻게 되는데, 이것이 옥상 설치물에 지배적으로 사용된 모양이다. 이처럼 단순한 모양을 현장에 배치할 때는 공간을 최대한 배려하면서 이 형태를 쌓거나, 나란히 배치하거나, 서로 한 데 뭉치게 함으로써 더 많은 사람들이 공간을 활용할 수 있도록 한다. 더불어 영사용 스크린이 그 중심을 이루게 한다.

COLOR ZONING

1 Roof and existing structures

2 Theatre

3 Platform and stairs

4 Scene 1

5 Scene 2

6 Theatrical tube(acoustics,luminiscence&applause)

7 Glass tube (Intemission)

8 Projection screen(performance)

Colour zoned plan showing Room sets

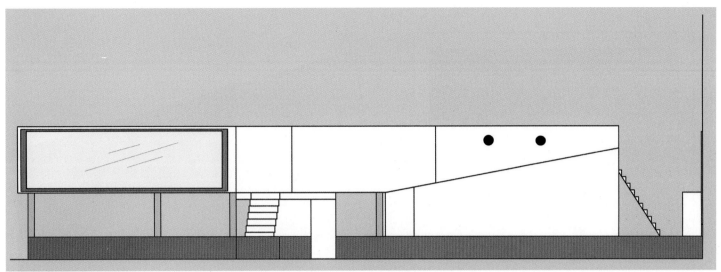

Elevation looking at the Theatre

Theatre: - The structure for sitting in and viewing presentations or videos.

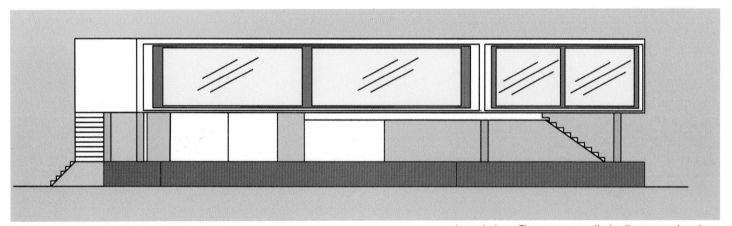

Elevation looking at Theatre through to Acoustics

Intermission: - The spaces or walks leading to certain points.

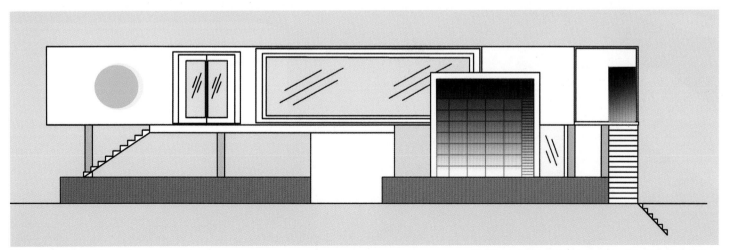

Elevation looking at Scene 2

Platform: - Most of the shapes are mounted on platforms.

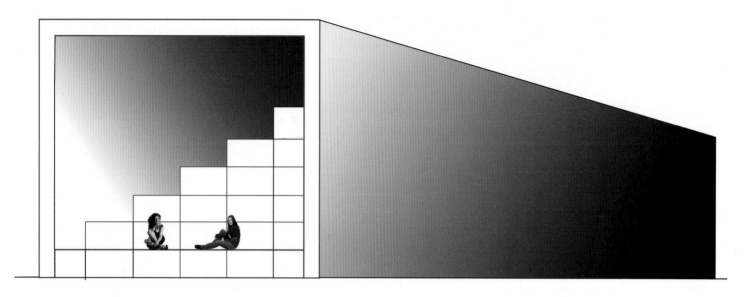

Burbstage is the combination of three sided cubes and cuboids (genealogy) mounted on platforms and the parallel use of theatrical terms or vocabulary to describe the parts of the proposed structure.

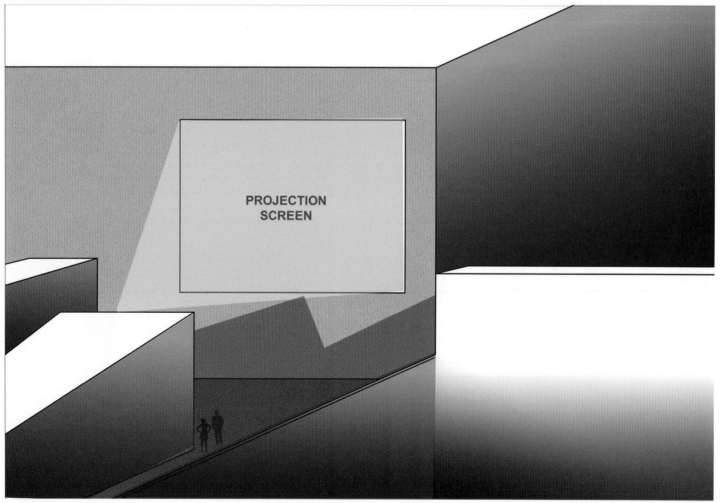

The Curtain Call is the fusion of their genealogy, history and passion for theatre.

Scene: - The viewing areas of the structure, overlooking Shoreditch and London.

Performance: - The projection screen in front of the theatre and the experience within the theatrical tube are representative of a performance.

Justin Melican
justinmelican@yahoo.com

Gallery TEN-CITY

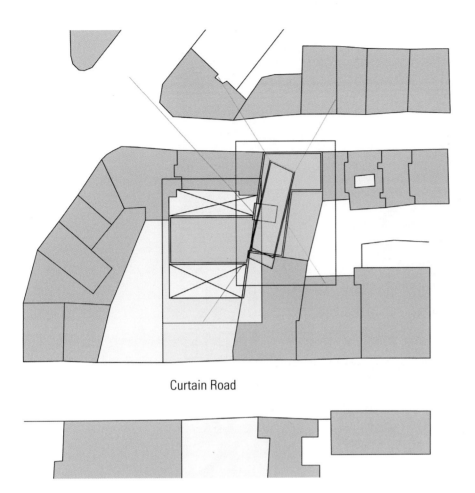

Curtain Road

The Story

In the last 10 years Shoreditch has seen a huge transformation; it has mutated from a rundown area to a hub for creative types, and in turn this has seen the emergence of great bars, pubs, galleries and restaurants.

The close knit streets with its former industrial context provide an area for those that want to step out of the norm and the corporate to imagine new businesses. Innovative creative houses have set up home using space imaginatively, sharing with like minded disciplines.

The narrowness of the streets encourages networking and enhances collective working practices. Warehouse constructions reveal their previous activity and function, reminding us of the busy people, the noise and the industry.

The buildings allow a maximum of light though to the interior, the detail in the window frames and brickwork show a sense of pride. The Interiors are simple and true the outer structure, revealing angles and curves. The wooden floors add softness and hollow noises that suggests occupancy.

The space is unattractive to the Multi- national corporations, there is no capacity for charging large rents which makes economically suitable for young and emerging practices to set up.

The rawness of the interior spaces, unadulterated by recent 20th century trends provide the perfect utilitarian and vacant habitat for the fresh ideas where pioneering and ground breaking professions materialize.

스토리

지난 10년 동안 쇼어디치는 엄청난 변화를 겪었다. 쇄락해가는 지역에서 창조적 비지니스의 허브로 탈바꿈하였고, 이로 인해 훌륭한 바, 펍, 갤러리, 그리고 레스토랑이 등장하게 되었다. 과거 공업지대로서의 배경을 지닌, 촘촘하게 얽힌 도로들은 평범함을 탈피하고자 하는 개인들과 새로운 비즈니스를 구상하는 업체들에게 훌륭한 공간을 마련해주었다. 창의적으로 공간을 활용한 혁신적이고 창조적인 주택들이 속속 들어서 비슷한 생각을 가진 사람들이 한 데 모여 살게 되었다.

도로가 비좁기 때문에 연대와 공동작업이 활성화되었다. 창고 건물들은 과거의 활동과 기능을 보여줌으로써, 분주한 사람들, 소음, 그리고 공장의 모습을 떠올리게 해준다.

건물들은 내부로 최대한의 재연광을 끌어들일 수 있도록 디자인되었고, 창틀과 벽돌에 나타난 섬세함을 통해 자긍심을 엿볼 수 있다. 나무로 된 바닥은 부드러운 느낌을 자아내고, 걸을 때마다 소리가 나기 때문에 누군가가 사용하고 있는 공간임을 쉽게 알 수 있게 해준다.

이 공간은 다국적 기업 등에게는 그리 매력적이지 않을 뿐더러, 높은 임대료를 책정할 여지도 거의 없기 때문에 패기만만한 신규사업자들에게 적합하다 할 수 있다.

최근의 경향이 전혀 반영되지 않은 날것 같은 느낌의 내부 공간은 친취적이고 혁신적인 사업이 뿌리내릴 수 있는 완벽하고도 실용적인 장소를 마련해준다.

Sketch showing the towering linear lines of the encroaching city

There's a war going on in Shoreditch, as the bold new high rises start to advance towards it, will we lose the essence of what Shoreditch is to corporation? You can literally see the waves of industry lapping at the shores of creativity.

Ten-city Gallery is an example of how new architecture can merge with the existing surroundings without compromising any of the current structures. By suspending the gallery on cables none of the current roof top is lost, the structure just floats among the buildings in whatever space is available.

The tension in the cables reflects the tension be-tween the two areas. The planting of wild flowers and grass on the roof top replaces the footprint of the building with what was originally there. I believe the gallery is a look into the future showing us how we can use hidden spaces.

새로운 고층건물들이 들어서기 시작하면서 쇼어디치에서는 현재 전쟁이 진행중이다. 우리는 대기업에게 쇼어디치의 본질을 내주게 될 것인가? 여러분은 말 그대로 산업의 물결이 창의성이라는 해변으로 다가와 찰싹거리는 모습을 목격하게 될 것이다.

텐 시티 갤러리는 어떻게 새로운 건축물이 현재의 구조를 전혀 헤치지 않고 기존의 환경에 조화롭게 녹아들 수 있는지를 보여주는 좋은 사례가 될 것이다. 갤러리를 케이블에 메달아 놓는 방식을 통해 기존 옥상 공간의 손실을 막았고, 공간이 있는 곳이라면 어디든 구조물이 건물 사이를 떠다닐 수 있다. 케이블들간의 팽팽한 긴장감은 두 지역 사이의 긴장감을 표현한다. 옥상에 야생화 및 야생초를 심어서 기존 건물의 흔적을 상쇄시킨다. 나는 갤러리를 통해 향후 우리가 숨겨진 공간을 활용할 수 있는 방안을 모색할 수 있으리라 확신한다.

Looking into the inner void of Burbage House

Looking at one of the three platforms

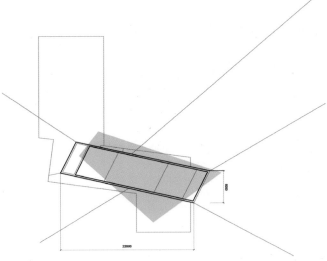
Plan high-lighting site lines

Behind Burbage House, the escape stair winds up the internal void improvised roof lights add light to the deep lower spaces of the building. At the top is a series of rooftop platforms that articulate over three buildings.

Ideas for planning, Taking ideas from the established lower framework, plan forms were sketched across the three areas, making one structure. The vertical and horizontal elements baffle the light and provide linear patterns across the whole of the space.

버비지 하우스 뒷쪽으로는 탈출용 비상계단이 있고, 옥상에 설치된 조명이 건물의 후미진 낮은 공간까지 빛을 비추고 있다. 꼭대기에는 일련의 옥상 플랫폼이 설치되어 3동의 다른 건물로 연결된다.

하위 프레임워크를 기초로 아이디어를 얻고, 세 부분으로 나뉘어진 계획안을 스케치해서, 하나의 전체 구조물을 만들어낸다. 수평과 수직으로 배치된 재료들이 빛을 가로막고, 선형의 패턴을 공간 전체에 만들어낸다.

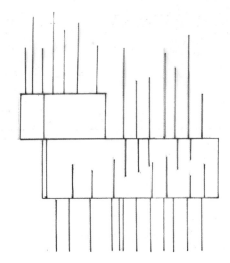
Elevation showing a right angled configuration

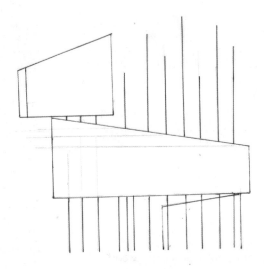
Section looking at the two levels

Modelling using Foam Board and wooden spills adding coloured filters to show the initial volumes of space.

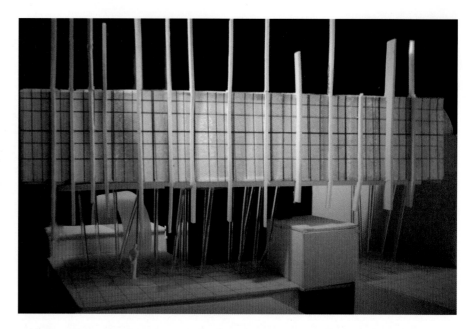
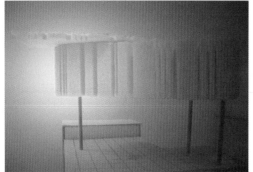
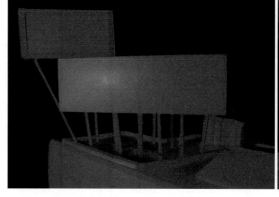
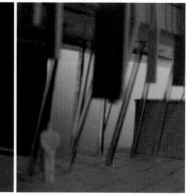
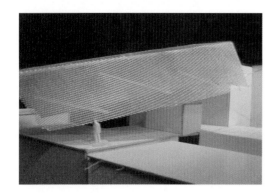

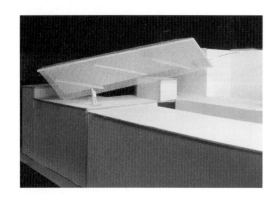

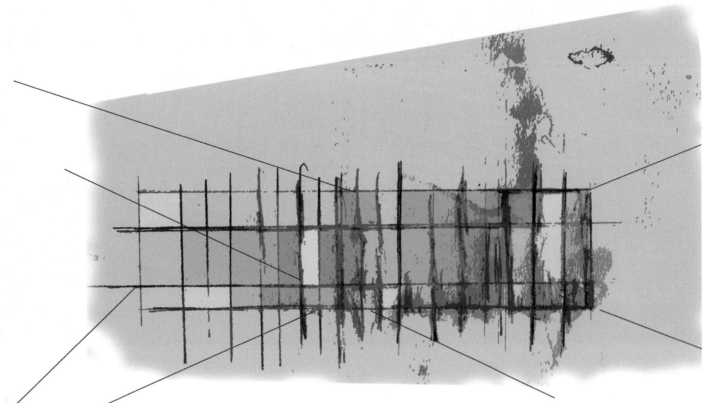

Two Renderings showing the articulation of linear elements of the rooftop construction. The tension lines linking to views and attaching invisibly to the buidings that surround both Curtain Road and Charlotte Road.

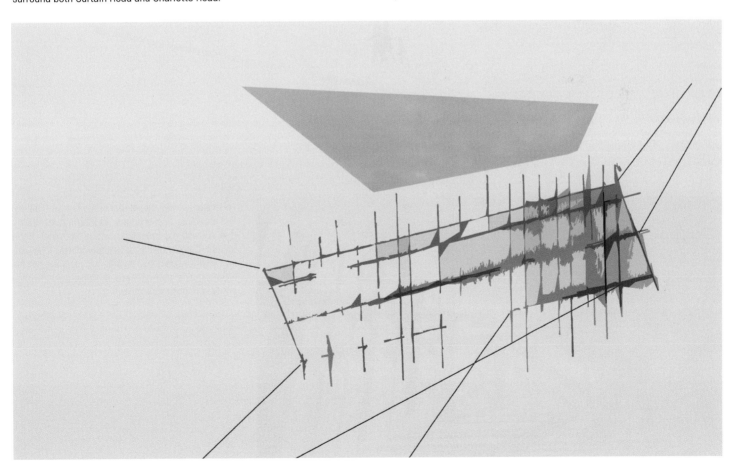

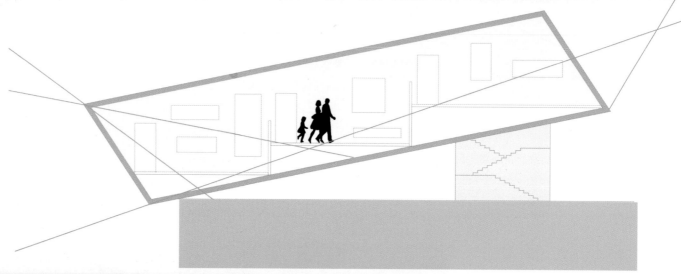

This Visual shows the interior of the Ten City Gallery. The steel intersections have an intermittent rhythm that adds to the irregular sizing in the older architecture. Compressed shapes situated next to expansive forms, economical and laconic planning but the inconsistency making each building an individual entity within the collective.

3D computer generated drawing of the cantilevered Ten City Gallery

As the lights falls through the structure dynamic shadows fall on the inner walls and floors The structure is made from gold coloured metal both internally and externally. The interior shows a mottled pattern which in turn also represents the tension, showing the metal to be stretching apart.

The gold exterior is there to be bold and to stand out and reflect the sun off it. I want everyone around and especially the people in the towering offices to see this shinning beacon of creativity. The cables which will support the structure will be connected to surrounding buildings the tension in the cables and the tension felt by people passing underneath the structure on street level as they look up to see a building which looks to be sliding off the above roof top is a direct connection.

구조물을 따라 빛이 아래쪽으로 떨어지면 역동적인 그림자가 안쪽 벽과 바닥에 드리운다. 구조물은 내외부가 모두 황금색으로 칠해진 금속 소재로 제작된다. 인테리어에는 얼룩덜룩한 패턴이 사용되는데, 이 역시 긴장감을 나타내며, 금속이 제각각 뻗어나가는 모습을 보여준다.

황금색 외관은 눈길을 잡아끄는 강렬함을 보여주며, 햇빛을 반사한다. 나는 주변의 모든 사람들, 특히 고층의 사무실에서 일하는 사람들이, 이러한 창의력의 상징물을 볼 수 있기를 희망한다. 구조물을 지탱하는 케이블은 주변 건물에 연결될 것이며, 팽팽하게 연결된 케이블은 긴장감을 자아내게 될 것이며, 거리를 걸으며 구조물의 아래를 지나가게 될 사람들은 마치 옥상에서 굴러떨어지는 듯한 구조물의 모습을 바라보면서 또 다른 긴장감을 느끼게 될 것이다.

The shadows will move around the space throughout the day, stretching and contracting revealing a woven pattern of differing tensions

Kathrin Acker
Kathrin.Acker@gmx.net

NEXUS

NEXUS

LOCATION

The rooftop is located in the middle of the city, in one of the most up and coming areas in London, Shoreditch. Curtain Road is a mish mash of buildings from years of industrial development. The architecture spans several centuries but the majority of buildings belong to the early twentieth century when the Industrial Revolution was well established and production methods were fast flowing. The River Thames was the artery to receive the raw materials from around the world and indeed export them out reformed.

AREA AESTHETIC

Industrial buildings were simply built to provide space for production; their aesthetic wasn't important and so was left unrefined and quite basic. It's this openness and emphasis on this utility that makes the space and the environment so attractive to young and emerging practices. They have an honest appeal, untainted by the brands and politics of others. It follows as a consequence of this rawness that the rents are cheaper, which of course is the biggest draw to a young creative company.

BURBAGE HOUSE

Burbage House is home to many creative companies which include architects, interior designers, graphic designers and film makers. They all share the space on the rooftop. Through access via a metal escape staircase each floor can find a special private space high above the roads and the hubbub of London, away from the rat race, the congestion, out hearing distance from the tension of van drivers zooming by. This is such a luxury in such a relentless metropolis.

위치

옥상은 최근 런던 시내에서 가장 주목받고 있는 지역 가운데 하나인 쇼어디치에 위치하고 있다. 수십년간의 산업화로 인해 커튼 로드의 건물들은 뒤죽박죽이 되어버렸다. 수세기 전에 지어진 건축물부터 현대적인 건축물까지 그 종류가 매우 다양하지만, 대개는 산업혁명이 성숙단계에 접어들어 제조기술이 빠르게 퍼져나가던 20세기 초반에 지어진 건물들이다. 템스강은 전세계로부터 원자재를 수입해 이를 가공해 수출하는 통로로서의 역할을 담당했다.

지역의 심미적 특징

공장건물들은 단지 상품 생산을 위해 건축되었으며, 심미적 요소는 배제되었다. 따라서 딱히 세련되지도 않았고, 지극히 단순한 형태를 띄고 있다. 젊은이들과 새롭게 부상하는 업체들에게 매력적으로 다가온 것은 바로 이러한 개방성과 실용 중심의 사고인 것이다.

이 지역이 지닌 꾸미지 않은 있는 그대로의 모습 덕분에 임대료는 낮아졌고, 젊고 창의적인 기업에게는 이점이 가장 큰 매력으로 다가오게 되었다.

버비지 하우스

버비지 하우스에는 건축, 인테리어 디자이너, 그래픽 디자이너, 영화 제작자 등 창의적 업체/개인이 입주해있다. 이들은 옥상에 있는 공간을 공유한다. 철재 비상계단을 통해 각 층에서 직접 옥상에 접근할 수 있는데, 도심의 복잡함과 교통체증에서 멀찍이 떨어져, 런던 한복판에서 즐기는 여유로움은 진정한 호사라 말할 수 있다.

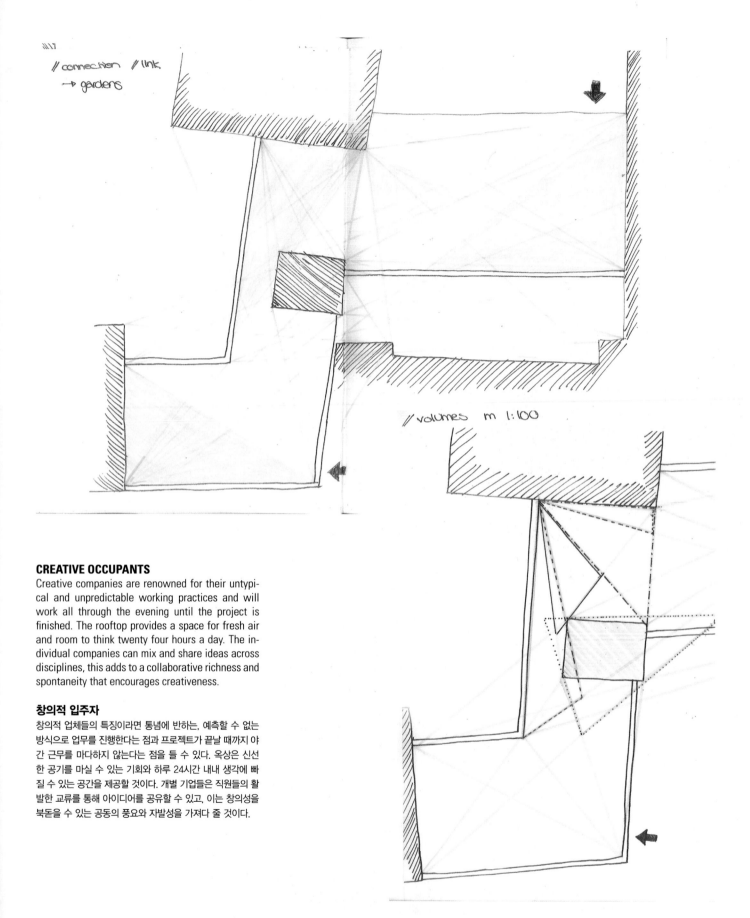

// connection // link
→ gardens

// volumes m 1:100

CREATIVE OCCUPANTS

Creative companies are renowned for their untypical and unpredictable working practices and will work all through the evening until the project is finished. The rooftop provides a space for fresh air and room to think twenty four hours a day. The individual companies can mix and share ideas across disciplines, this adds to a collaborative richness and spontaneity that encourages creativeness.

창의적 입주자

창의적 업체들의 특징이라면 통념에 반하는, 예측할 수 없는 방식으로 업무를 진행한다는 점과 프로젝트가 끝날 때까지 야간 근무를 마다하지 않는다는 점을 들 수 있다. 옥상은 신선한 공기를 마실 수 있는 기회와 하루 24시간 내내 생각에 빠질 수 있는 공간을 제공할 것이다. 개별 기업들은 직원들의 활발한 교류를 통해 아이디어를 공유할 수 있고, 이는 창의성을 북돋을 수 있는 공동의 풍요와 자발성을 가져다 줄 것이다.

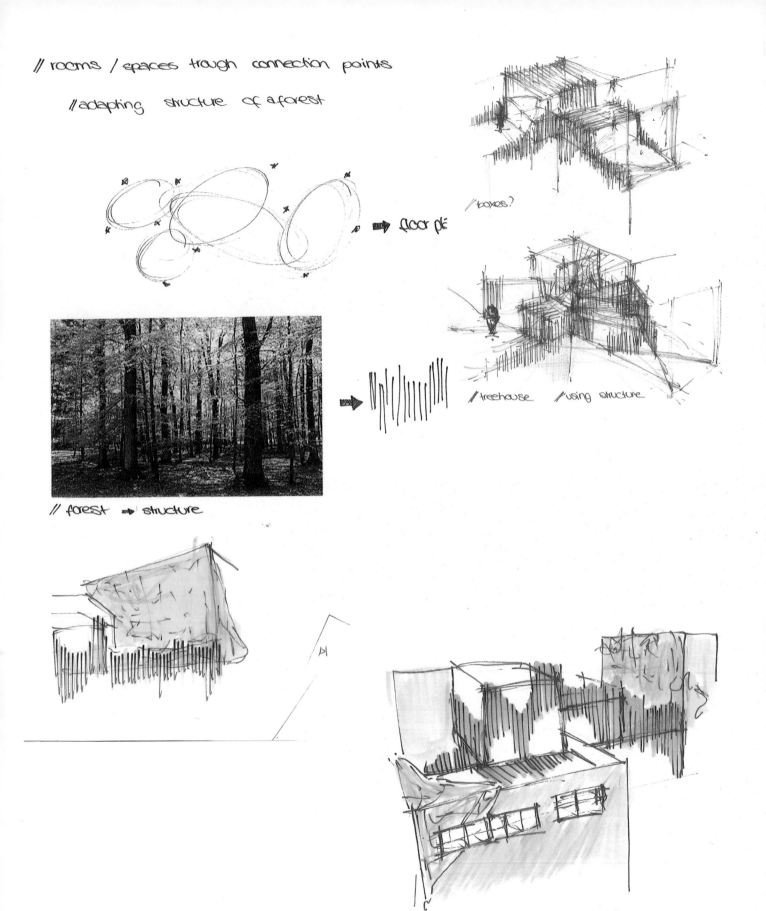

// rooms / spaces trough connection points

// adapting structure of a forest

➤ floor plan

// boxes?

// forest ➤ structure

// treehouse   // using structure

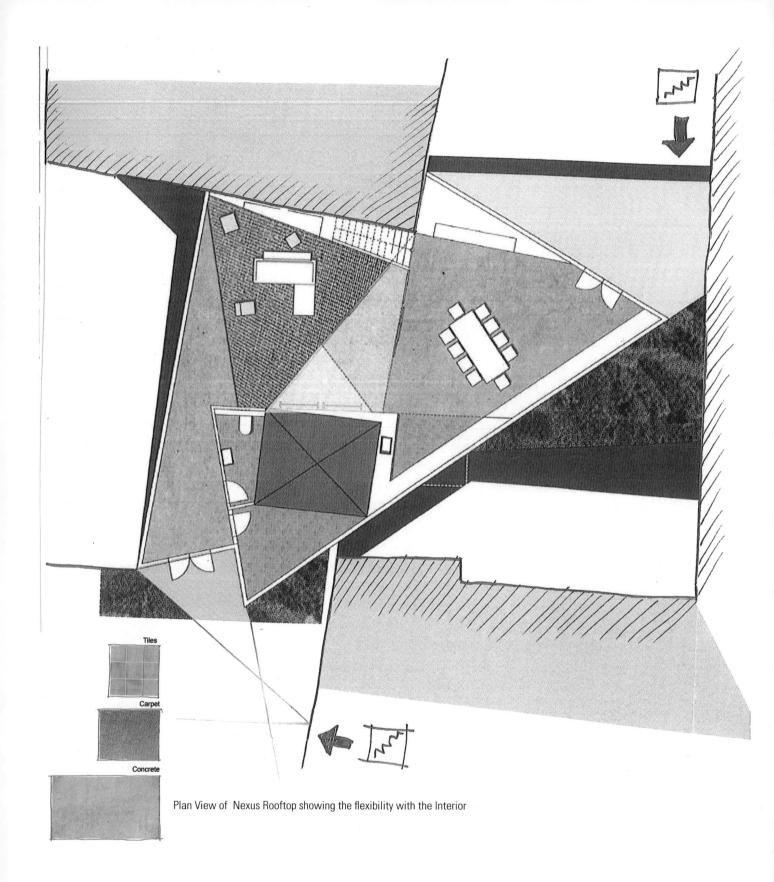

Tiles

Carpet

Concrete

Plan View of Nexus Rooftop showing the flexibility with the Interior

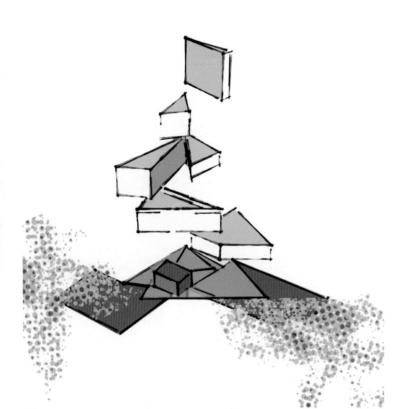

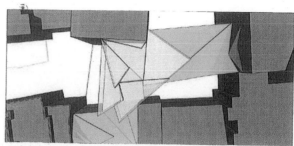

// different views without scale

// transparent top view

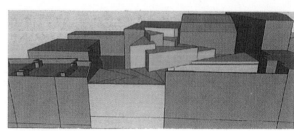

// east view
west

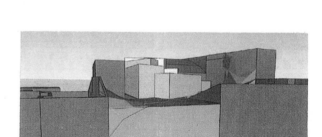

// east view, difference between vegetation
west   and building

// heights   m 1:100

6.50m

4.00m

5.80m   2.50m

3.70m

4.50m

// section

Diagnostic of the elements and forms showing the contrast between the soft planting and the sharp edges of the interlocking shapes.

DESIGN INTENTION

The design recognizes this precious open space and it's connectedness to clear thinking and exchange of ideas and wants to augment all these positive attributes by adding soft green space for relaxing and relaxing. A vertical garden is established on one of the two tall walled boundaries. The creeping plants will grow and crawl onto the site and also over the building facades and fall down onto the roads below. Thus allowing the passersby to realize that the roof is being used adds to the buildings mystique.

디자인 의도

디자인 의도는 옥상의 소중한 열린 공간을 활용해 맑은 사고와 아이디어의 교환을 이끌어 낼 수 있는 장소를 만드는 것이다. 이러한 긍정적 특성들을 살리기 위해 마음을 편안하게 해 주는 부드러운 녹색 지대를 조성한다. 옥상 둘레를 감싸는 두 개의 높은 벽들 가운데 한 곳에는 수직으로 정원을 설치할 것이다. 이곳에서는 덩굴식물들이 자라나 벽 전체를 덮는 한편, 건물의 정면까지 뻗어나가 아랫쪽 도로까지 계속 자라게 될 것이다. 근처를 지나는 행인들이 이 모습을 통해 옥상이 활발하게 사용된다는 점을 알게 될 것이며, 건물은 보다 신비롭게 느껴질 것이다.

DESIGN REFERENCE

The design takes reference from the existing buildings, their architecture, forms and materials and uses this language in an abstracted way to compose three spaces that articulate over the three different sections of the rooftop.

CONNECTIONS

The design endeavours to make connections to the surrounding buildings by line association and the scale used in choice of materials. The use of triangular shapes throws the eyeline across to witness the surroundings and differing perspectives and thus attaching the new to the old.

디자인 참고사항

디자인을 위해 기존 건물들의 건축법, 형태, 재료 등을 참조하였고, 이를 통합해 세부분으로 나뉘어진 옥상 공간 전체를 서로 연결하는 3개의 공간을 만들어냈다.

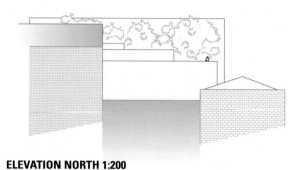

ELEVATION NORTH 1:200

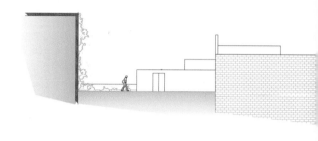

ELEVATION EAST 1:200

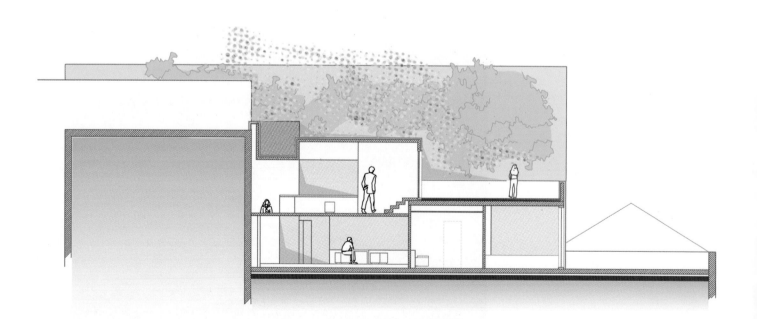

SECTION B-B 1:100

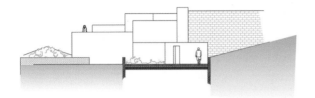

ELEVATION SOUTH 1:200

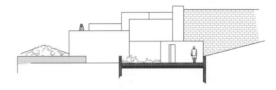

ELEVATION WEST 1:200

FLEXIBILITY

These design of this rooftop has created a flexible meeting point for the users of Burbage House both privately and perhaps more publically as they can host exhibitions and offer the space to a wider audience.

유용성

옥상에 대한 이러한 디자인은 버비지 하우스 입주자들에게 매우 유용한 만남의 장소를 마련해주었다. 개인적으로 이용할 수 있을 뿐만 아니라 전시회를 열거나 더 많은 관객들에게 공간을 개방함으로써 보다 공개적으로 활용할 수도 있을 것이다.

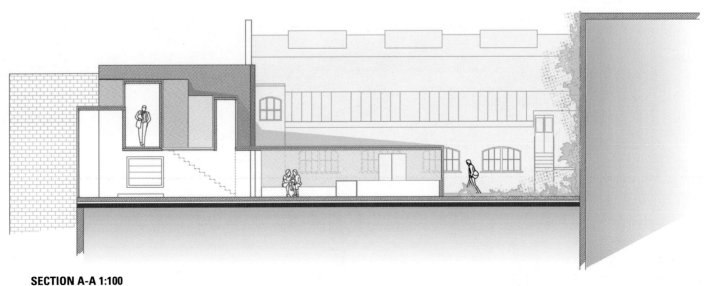

SECTION A-A 1:100

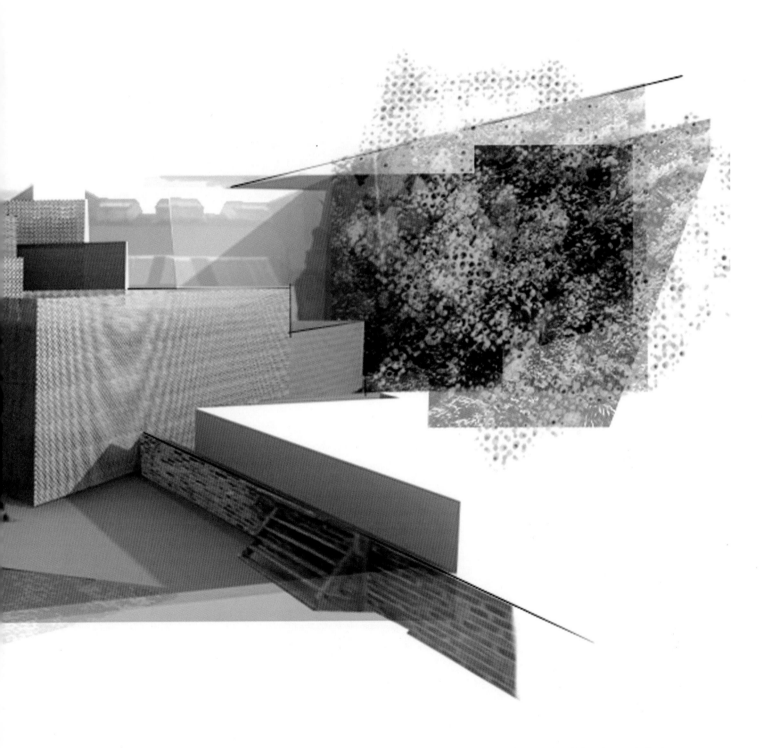

Perspective showing the relaxation space which looks out to Shorditch. The metal mesh facades nestle in the greenery.

Thinking space

Interior Perspective showing white empty space, furnished with oak and other natural materials.

FIRSTFLOOR 1:100

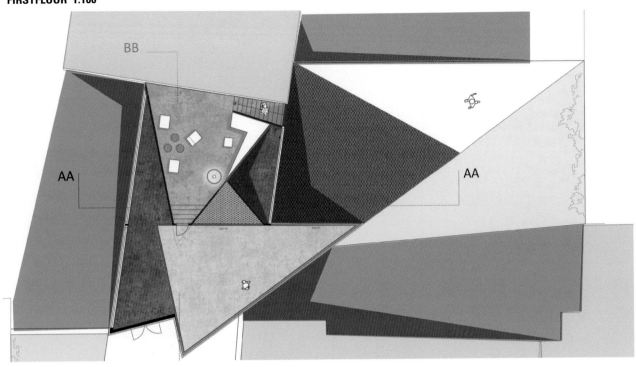

GROUNDFLOOR 1:100

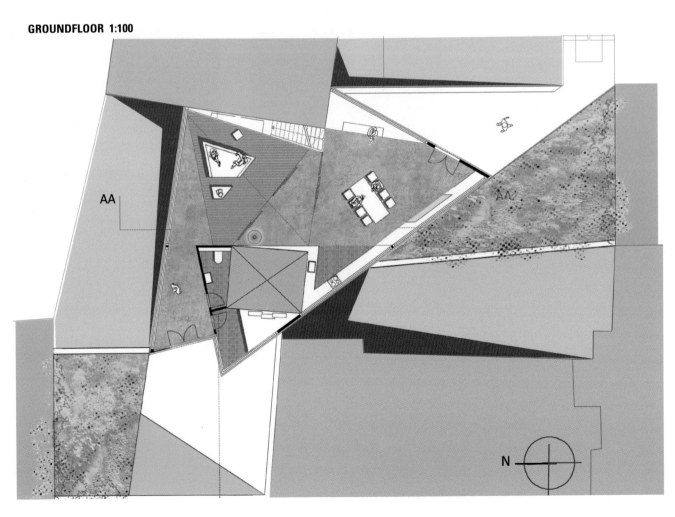

N

A higher perspective of a communal area, realxed and informal.

Marion Christiaens
marion.christiaens@gmail.com

Urban Thread

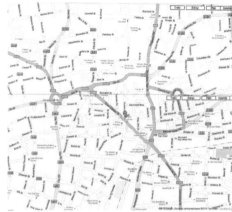

Context

The site of the roof top Hidden Space project is in Shoreditch in the East End of London.
The access to the Burbage House is from Curtain Road. Winding through and up the building you reach the roof top, shared by five creative design companies. The shared space currently acts as a breakout and social space. The aim is to continue and embrace the community, connection and co-operations that exist by devising areas for exhibition, social interaction and contemplation.

이번 프로젝트를 진행할 장소는 런던 이스트 엔드의 쇼어디치 지역이다.
버비지 하우스의 출입구는 커튼 로드를 면하고 있다. 건물을 따라 올라가면 옥상에 다다르게 되는데, 이곳에 입주한 다섯 곳의 창의적 기업이 이 공간을 공유하고 있다. 옥상은 현재 휴식과 사교의 공간으로 활용되고 있다. 프로젝트의 목표는 기존의 커뮤니티, 연계, 그리고 협동을 그대로 유지하면서, 전시와 사교, 그리고 사색의 공간을 만들어내는 것이다.

Urban Network

Through the articulation of the Urban Network, mapping the connections across the area of architectures, creativity, and materiality, schizophrenic patterns of juxtaposing massing and streets, form the starting point for the hidden rooftop space.

도시 네트워크

쇼어디치의 주요 도로들이 핵심 구조를 형성한다. 더불어 도로의 차선들과 골목들은 내부 구조를 만들어낸다. 사람들과 이들의 활동은 인프라가 존재하는 이유가 된다. 작은 정착지들로부터 서서히 도시가 성장한다. 긴 세월과 많은 세대를 통해 성장과 디테일이 보태진다. 언뜻 눈치채기 어려운 끊임없는 움직임이 복잡한 사회 구조로 발전해나간다.

Looking at the Rooftop from Burbage House.

Image shows threads crossed across a volume reflecting on the floor.

The Major routes through Shoreditch provide the main structure. The streets lanes and alletways then add the interior structural forms. People and function give the infrastructure reason to exist. A city builds from the smallest of settlements. Many years and generations add to the growth and detail. This flux seemingly un-noticeable evolves into a very intrincate society of structure.

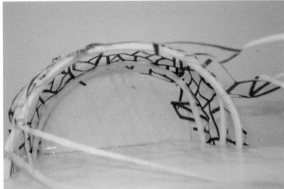

Model making exploration. Each techniques adds characteristics to the space. The width of the thread, its roundedness is important The tension of the thread implies it's connected attachments therefore reinforcing the the idea behind the concept. It's essential that rythm has the same propoerties as the streets in the Shoreditch area. Again tense, assymetrical.

AA

BB

The Rooftop

The manipulation of patterns and links within the network, forms both private and public space, through, under, within and around a wooden structure. The structure connects guides and punctuates the space, encouraging meeting points and places of solitude, to create some head space from the office environment. The framing of view and vista provides an outward perspective on the context of the area, however the flexible relationship within the space and it's surrounding allows the user to take make their own choice of how it is inhabited.

옥상

패턴의 변용과 네트워크 내부의 연계를 통해 목조 구조물의 아래에, 내부에, 그리고주변에 사적인 동시에 공적인 공간이 탄생한다. 이 구조물은 만남의 장소인 동시에 고독의 공간이며, 사무실 환경을 벗어날 수 있는 열린 공간이 된다.
옥상은 사교 목적의 행사나 전시회에 이용할 수 있고, 입구쪽에 위치한 9.5미터 높이의 벽을 이용해 프레젠테이션 용도로도 활용할 수 있다. 패션쇼나 영화 상영 역시 건물에 입주한 업체들이 이 공간을 활용할 수 있는 좋은 사례가 될 것이다.

Provocate the Meeting

Visual show the intrincate layering of structurl framework giving a ceiling and adding opportunities to sit and divide spaces.

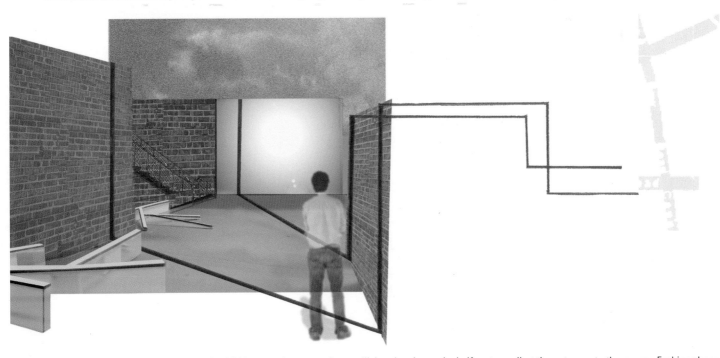

The rooftop can be used for social events and exhibitions, and presentations, utilising the nine and a half metre wall at the entrance to the space. Fashion show, films screenings are some of the examples that those within the building can utilise and exploit the space. When the screen is not in use an artist's light instillation can create, a textured ambiance to the space.

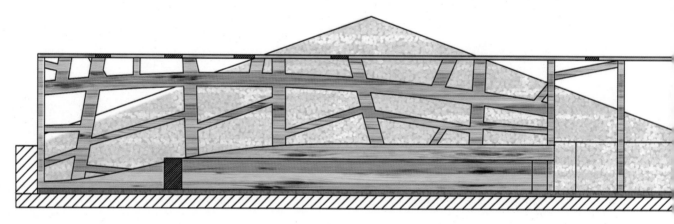

SECTION A-A

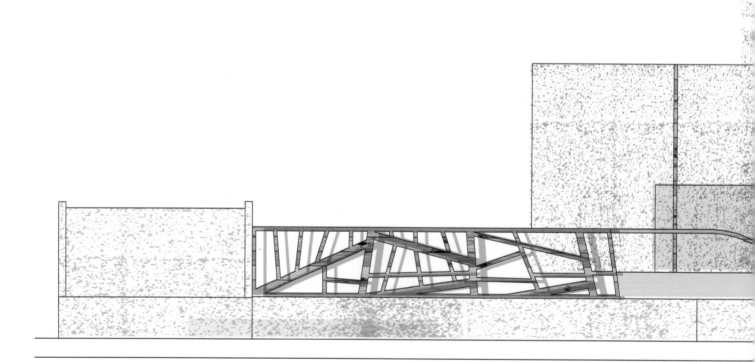

SECTION B-B

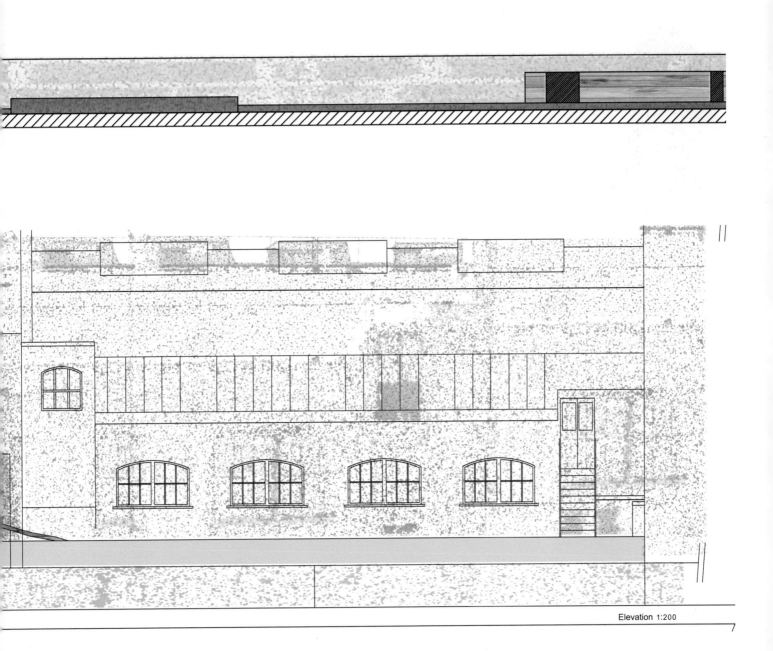

Elevation 1:200

Pimprapai Thapthimsuk
baimint08@hotmail.com

TIMELINE

The ground was leased to Burbage in 1576, and formed part of the precinct of the Priory of St. John the Baptist, Holywell, just outside the City's boundary, within which dramatic performances were strictly prohibited.
From London's Coats of Arms and the Stories They Tell –

1576

COWPER ST

Several streets of houses were built upon the site, as was also Holywell Mount Chapel.
From London and Middlesex

1787

The chartered Gas-Light Company was the first established in London, having been incorporated in 1812.
From Leigh's new picture of London, or, A view of the political, religious, …

1812

NEW INN YA

Lord Strange's men opened on February 19, 1592, a third London theatre, called the Rose, which Philip Henslowe, the speculative.
From A Life of William Shakespeare

1592

STREET

A stained-glass window erected in St. James's Church, Curtain Road, in 1886, exactly 300 years after Shakespeare first came to London, commemorates his somewhat conjectural connection with these theatres.
From Full text of "London and its environs"

1886

Middlesex County Records under the date February 21, 1627.I It is merely a passing reference to "the common shore near the Curtain playhouse," yet it is significant as indicating that the building was then still standing.
From Shakespearean Playhouses

1627

CURTAIN ROAD

The London County Council whose conclusion that The Theatre was within a few yards of the Curtain road School is uplifted by a good deal of documentary evidence ground.
From "the Theater". Of London's First Play House.

1915

The New Inn Yard and King John's Court marked remains of the Priory of St. John the Baptist, founded by a Bishop of London for black nuns of the Order of St. Benedict.
From Full text of "Notes and queries"

1745

In January 1995, Chris Herbert booked the group's first professional songwriting session with the producers at the Strong room in Curtain Road, East London.
From Wannabe (song) - Wikipedia, the free encyclopedia

1995

Finsbury Square

Dennis

Panoramic view from the Rooftop of Burbage House, 83 Curtain Road, London EC2

Timeline Concept

The Timeline concept endeavours fuse the spirit of space of past, present and future of Curtain Road. It creates new space for the creative companies who work within Burbage House and structure that both relaxes and inspires.

컨셉트

타임라인의 컨셉트는 커튼 로드의 과거, 현재, 그리고 미래의 정신을 한 데 섞는 것이다. 이를 통해 버비지 하우스에 입주한 창의적 업체들을 위한 새로운 공간, 휴식과 영감이 교차하는 공간을 만드는 것이다.

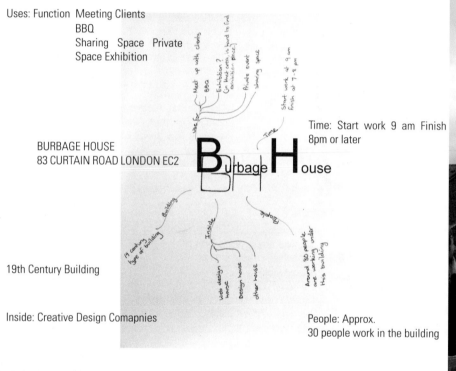

Uses: Function Meeting Clients
BBQ
Sharing Space Private
Space Exhibition

BURBAGE HOUSE
83 CURTAIN ROAD LONDON EC2

Time: Start work 9 am Finish
8pm or later

19th Century Building

Inside: Creative Design Comapnies

People: Approx.
30 people work in the building

Inspiration is taken from the surrounding panorama showing the close knit city planning where the individuality of each structure is apparent, each with their own function and timeline. The squeezing and squashing shapes the skyline as it lights and drops, with a rusticated and staggered appearance.
The interlocking shapes reveal gaps allowing for light and air to squeeze through and at the same time allowing glimpses of what's behind, in and beyond. A timeline of activity and need relating to economics, politics and London's place in the world. The structural history.
It's this sense of time, affluence, work and integrated activity that makes London such a vibrant and ever changing place to be in. The Timeline design wants to reflect this in the reflective and fluid materials but integrates this mirroring with the structural memory of the cityscape.

촘촘히 차여진 도시의 모습이 마치 파노라마처럼 펼쳐진 주변환경으로부터 영감을 얻었는데, 각각의 구조물들은 뚜렷한 개성을 갖고 있을 뿐만 아니라, 그 나름의 기능과 타임라인을 지니고 있다. 찌그러지고 튀어오르는 등 다양한 모습이 스카이라인을 수놓고 있다.
서로 맞물리는 형체들은 빛과 공기가 비집고 들어올 수 있는 틈새를 드러내는 동시에 그 배후에, 안에, 그리고 아래에 무엇이 자리잡고 있는지 넌지시 보여준다.

런던은 경제, 정치, 그리고 세계사에 있어서 매우 중요한 위치를 차지하고 있으며, 끊임없이 변화하는 활기찬 도시이다. 타임라인의 디자인은 이러한 사실을 반영하기 위해 빛을 반사하는 부드러운 소재를 사용하였고, 도시경관에 대한 구조적 기억을 이와 결합시키고자 한다.

The semi-exposed roof design is unique as it is built with unevenly planed walnut beams with a 500cm gap allowing the natural light and air into interior space.

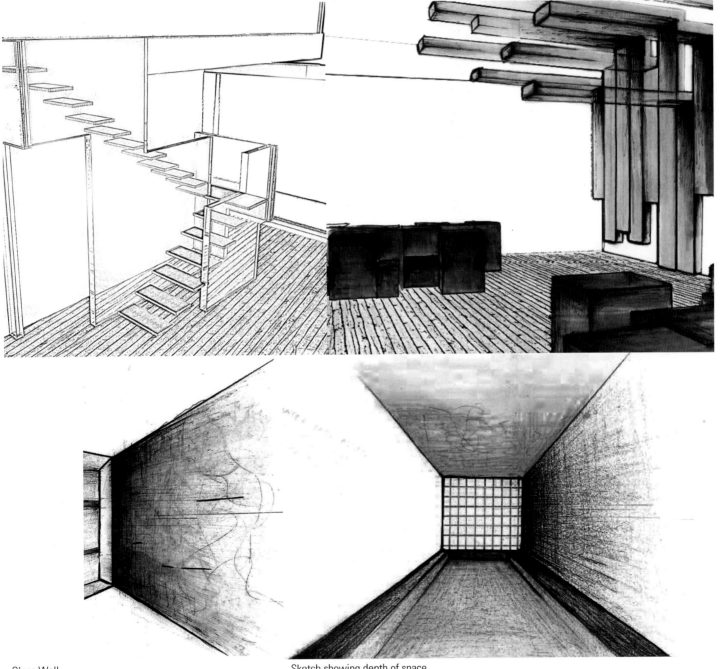

Glass Wall

Sketch showing depth of space

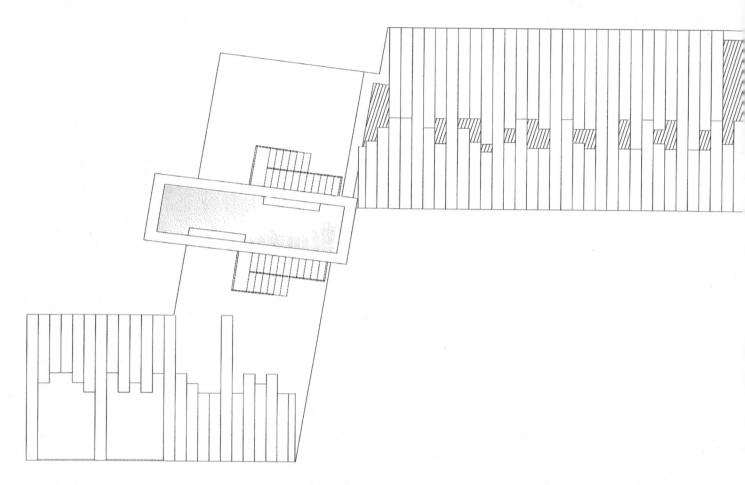

Plan outling the Intermittent Walnut Beams.

Front Elevation showing the Reflective Area, Sunburst coloured staircase and space to collect within The Private exhibition area is raised to distinguish it from the rest of the space.

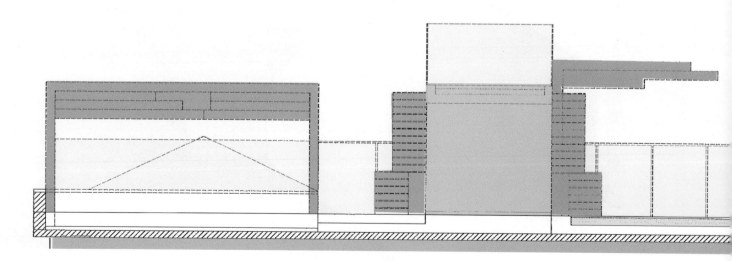

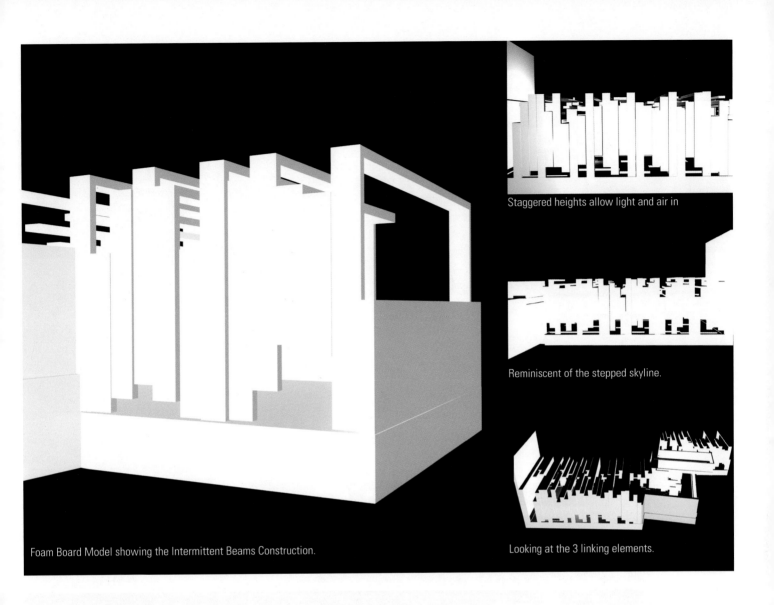

Foam Board Model showing the Intermittent Beams Construction.

Staggered heights allow light and air in

Reminiscent of the stepped skyline.

Looking at the 3 linking elements.

Reflective Area

Seat floating on Infinity Pool.

Entrance

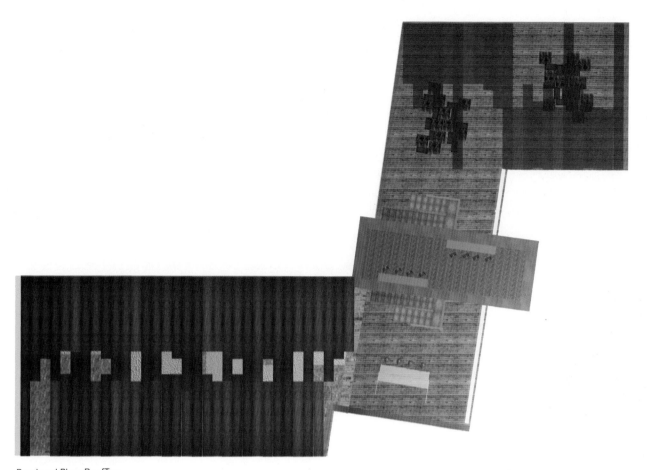

Rendered Plan: RoofTop

A rippling horizontal plane of water reflects the natural light and forms a magnificent moving of shadow along the walkway to create a shimmering feel to this space. The space is topped with a rough textured sustainable glass surface which mirrors the path.
The bar area represents current architectural design and contains a curtain wall of water revealing inter-mittingly and connecting through to the relaxation space.

잔물결이 일렁이는 수면은 햇빛을 반사하고, 통로에는 놀랄 만큼 아름다운 그림자의 움직임이 펼쳐진다. 이를 통해 공간은 일렁이는 듯한 느낌을 얻게 된다. 공간의 윗부분은 거친 질감의 지속가능한 유리로 덮히게 되는데, 유리 표면에는 통로가 반사된다.
바 공간에는 현대적 건축 디자인이 적용되며, 커튼형 물의 벽이 설치되어 이곳과 연결된 휴식 공간의 모습을 살짝 드러낸다.

Staircase to Infinity Pool

Infinity Pool

Ruth Samuel
ruthsamuel@btinternet.com

The Collective Soul

This area has more graffiti art work then any other single place in London from brick lane to Angel Market, and some of those were done by the infamous and elusive Banksy, some are replaced while others seem protected.

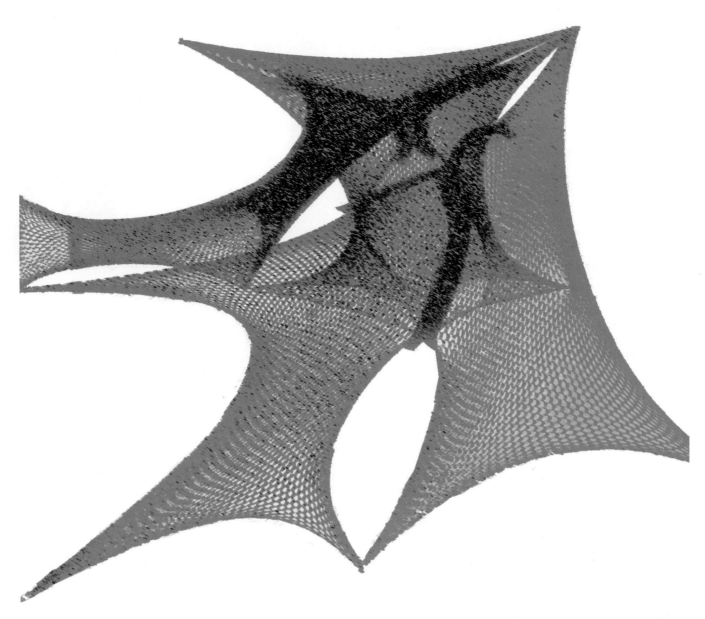

Curtain Road Shoreditch

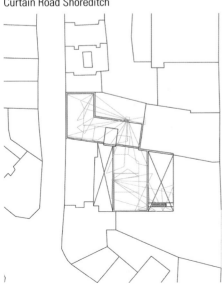

Burbage House

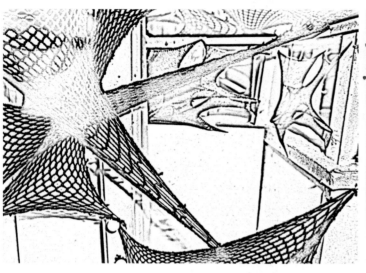

THE AREA

The area, steeps in the richness of the past with its architectural styles of old can assume high status as it is mentioned in the doomsday book. It holds on to its values, ethos and brand which are now protected by a conservation order.

Today, the area can be described as vibrant, transient and inclusive while maintaining its legacy.

Despite my reflecting there is still something missing. Burbage House display nothing of this on the building, there is no drama, no showing off of the nature of the people inside, also there are no signs of interaction between the people inside and the community, even those just across the road.

지역

이 지역은 오래된 건축 스타일과 더불어 풍요로운 과거의 모습을 보여준다. 지금도 여전히 이 지역만의 가치, 기풍, 그리고 개성을 지켜나가고 있다.

오늘날 이 지역은 전통을 고수하는 동시에 생동감있고, 계속 변화하며, 매우 포괄적인 공간이라 표현할 수 있다.

이러한 나의 평가에도 불구하고, 여전히 허전한 구석이 있다. 버비지 하우스가 건물을 통해 보여주는 것은 아무것도 없다. 드라마도 없고, 내부에서 생활하는 사람들의 성격을 보여주지도 못 한다. 또한 건물 내부 사람들과 커뮤니티 사이의 교류, 하물며 바로 길건너편 사람들과의 교류 조차 겉으로 드러나지 않는다.

THE CHALLENGE

The challenge however is bringing together the old and new, to push the boundaries I will attempt to put the soul back into Burbage house though intelligent contemporary architectural designing while being aware of the sensitivity of the area.

도전과제

나는 오래된 것과 새로운 것을 한 데 어우러지게 함으로써 공간의 경계를 확장하고자 한다. 이 지역의 정서를 충분히 이해하는 동시에 현대적 건축 디자인을 적용함으로써 버비지 하우스에 다시금 영혼을 불어넣고자 한다.

THE DESIGN

The design will emerge from the work space of the creative energy within the building to transform the roof into an environment of positive thinking, to enhance, to be dramatic, gleaming and distinctive, inviting to both visitors and tenants, then cascade down to mingle with the other creative's, such as the graffiti artist, of South Shoreditch putting on a performance both day and night. Working with the Kinetics the space is intended to be as flexible as possible, a gallery one week an office the next. Therefore materials such as glass, stainless steel and fabric will be used in its construction, which will follow the rhythm of the structural angles of the roofs around, I will bear in mind the environmental issues and where possible to use recycled and renewable materials while being aware of the limitations.

디자인

디자인은 건물 내부의 창의적 에너지를 토대로 그 모습을 드러내 옥상을 긍정적 사고, 발전, 드라마틱함, 찬란함, 그리고 독특함 등이 가득한 공간으로 변모시킨다. 방문자와 입주자 모두를 끌어들인 뒤 그 영역을 확장해 또다른 창조적 영혼들, 예를 들면 사우스 쇼어디치의 그래피티 아티스트 등과 손잡고 밤낮으로 멋진 퍼포먼스를 선보인다. 공간 활용에 있어서는 최대한의 융통성을 발휘해, 한 주는 갤러리로, 다른 한 주는 사무실로 사용하는 식이 될 것이다. 따라서 유리, 스테인레스 스틸, 섬유 등의 소재가 건축에 사용될 것이며, 옥상 구조의 앵글이 보여주는 리듬을 최대한 살리는 방향으로 진행될 것이다. 나는 환경과 관련된 사안들을 염두에 둘 것이며, 재활용되었거나 재생가능한 재료를 사용할 수 있을 경우 적극 활용할 것이다. 물론 그 한계에 대한 인식도 빠뜨리지 않을 것이다.

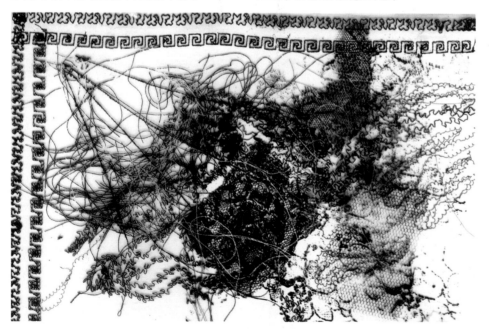

The gallery/office is made versatile by the use of sliding partitions which allows the users to determine the size and position of the room needed, these partitions are also made from glass which can be folded back against a wall when not in use; the glass is treated with a colour and a design for privacy with a further stretch design forming an arc, the designs on all glass window and partitions, and those forming the arc is similar to the stretched design on the roof.

The stretch designs are made from "fish net" tights or stocking (see figs 1) window designs was done using the corner of a room the fabric stretches from one side to the other, the design interact with the lift as well as running down the front of the building, however here it is painted on to the wall. (See fig 2).

The pattern that is thrown across the interior space by the design adds to the atmospheric mood I want for this space.

이 공간은 갤러리와 사무실 등 다용도로 사용될 수 있는데, 이는 사용자가 임의로 특정 공간의 위치와 크기를 정할 수 있도록 해주는 슬라이딩 파티션 덕분이다. 또한 이러한 파티션은 유리로 제작되는데, 사용하지 않을 때에는 벽쪽으로 접어놓을 수도 있다. 유리에는 프라이버시를 위해 색깔과 별도의 디자인을 가미하고 이후 스트래치 디자인을 통해 원호 모양을 만들게 된다. 모든 유리창과 파티션에 사용된 디자인은 지붕의 스트래치 디자인과 조화를 이루게 된다.

스트래치 디자인은 소위 "망사스타킹"에서 그 영감을 얻었고, 창문 디자인은 직물이 한쪽 끝에서 다른쪽 끝으로 뻗어나가는 한 공간의 모서리 부분을 이용해 이루어졌다.

디자인에 의해 내부 공간 전체에 퍼져있는 패턴은 내가 이 공간에 기대했던 전반적인 분위기를 한층 향상시킨다.

I shall also work toward reducing energy consumption and lower running cost via the installation of wind or solar power, this will be very important here as this project will have some led lighting throughout the night on the staircase and lift, the lift will move up and down during the night, remotely changing colour as it goes.

It should be fully inclusive providing something for everyone that could be used by all creative's within the area this could add real value to the collectives by networking and collaborating with people using the space. The area should be in Constance use even during the weekends, on Sundays the collectives can introduce their family to the space, work and colleagues over brunch while doing something creative that could possibly be displayed in the gallery.

The gallery could benefit all creative's as a place to promote, show or sell their work.

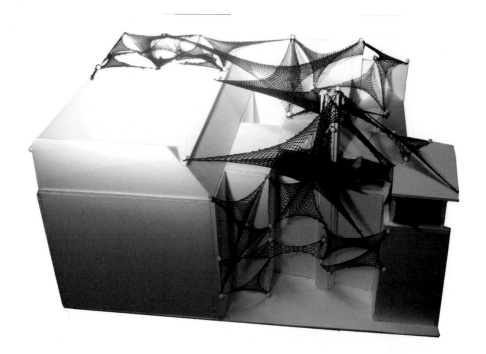

나는 또한 풍력 또는 태양광 발전을 이용해 에너지 소비와 유지비용을 줄이고자 한다. 이 프로젝트는 야간에 계단과 승강기 등에 LED조명을 사용할 것이므로 이 점은 매우 중요하다고 할 수 있다. 승강기는 야간에 조금씩 그 색깔을 바꿔가며 운행될 것이다.

옥상은 모자람이 없는 공간으로 꾸며져, 지역에 거주하는 모든 창의적 사람들이 기분 좋게 이용할 수 있어야 한다. 공간을 사용하는 사람들 사이의 협력과 네트워킹이야 말로 진정 공간의 가치를 높여주는 일이 될 것이다. 공간은 심지어 주말 동안에도 꾸준히 사용될 수 있다. 가족들을 데리고와서 옥상을 구경시켜 줄 수도 있고, 동료들과 브런치를 나누며 창조적인 작업을 벌일 수도 있을 것이다. 이 때 만들어진 작품이 갤러리에 전시될 수도 있다.

갤러리는 모든 창의적인 사람들에게 혜택을 줄 수 있는데, 이곳에서 그들의 작품을 홍보하고, 전시하거나, 판매할 수 있을 것이다.

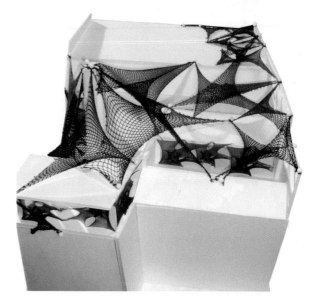

The History of Burbage House

The house got its name from the Burbage family, who were in the theatre business during the mid 1500s, the first theatre called 'The Theatre' was across the street from the now Burbage House was managed by this family, passing from father to son, and it played host to Shakespeare and his plays such as Romeo Juliet, two years later they opened another theatre, The Curtain, but the history goes back further.

History tells us that at some stage the Burbage family moved the building and re-built across the River Thames, this time they named it The Globe. Today the Globe has been re built again to look as it did then. As the area is within walking distance to the River Thames, London Bridge and The Tower of London it has always been central to the heart of London, as far back as the Doomsday book as it is mentioned there it became a good place for artisan who did not or could not set up business with in the so called square mile, from weavers to cabinet maker set up in the area and with them came apprentices who wish to learn the trade.

Today the area has a number of galleries, artist studios, exhibition, museums and fashion school, during the day, at night it come to life with its bars, restaurants, comedy clubs, small theatre, Indi cinemas, and night clubs, it ranks

버비지 하우스의 역사

버비지 하우스라는 명칭은 1500년대 중반 연극계에 몸담았던 버비지 가문에서 유래했다. 런던 최초의 극장인 "더 시어터"가 지금의 버비지 하우스 건너편에 있었다. 현재 버비지 하우스는 이 집안에 의해 대물림되며 관리되고 있다. 극장에서는 로미오와 줄리엣 같은 셰익스피어의 작품이 공연되었으며, 2년 뒤 또다른 극장인 "더 커튼"이 개관하였다. 하지만 역사는 여기에서 멈추지 않는다.

역사에 따르면 버비지 가문은 건물을 옮겨 다시 짓게 되는데, 이 때 템스강변에 세워진 극장이 바로 "더 글로브"였다. 오늘날 글로브는 옛 모습 그대로 복원되어 있다. 템스강, 런던 브릿지, 그리고 런던탑까지 도보로 이동할 수 있기 때문

에 이 지역은 언제나 런던의 중심이었고, 둠스데이북에 언급된 것으로 미루어 그 역사가 한참 전까지로 거슬러 올라가리라 짐작할 수 있다.

소위 스퀘어 마일이라 불리던 런던의 번화가에 가게를 차릴 수 없었던 장인들에게 이곳은 훌륭한 장소가 되었다. 직공에서 가구공에 이르기까지 다양한 직종의 장인들이 가게를 차렸고, 이들을 따라 많은 도제들이 기술을 배우기 위해 몰려들었다. 오늘날 이 지역에는 수많은 갤러리, 작업실, 전시장, 박물관, 그리고 패션 학교 등이 들어서 있다. 밤이 되면 바, 레스토랑, 코미디 클럽, 소규모 연극 공연장, 인디 영화 상영관, 나이트클럽 등이 이 지역에 생기를 불어넣기 시작한다.

Burbage House

Is situated between Old Street and Great Western Street, a stone throw away is Bishops gate, Moorgate and Liverpool Street Station with there iconic towers.

The other four buildings stands slightly lower, two of them are on either side of Burbage House, the other two are to the rear of Burbage House, the front of these two building opens on to Charlotte Street. The linking of these building is providing them all with fire exits, by crossing the roof then entering another building.

The people who are presently using the building are all so different, ranging from a finance company called Action Visas, Bona Fide Studios who are into music, Diverse Interior Architect LTD then Lucky bite who describes themselves on the web as "design and development of new product" and many more. All this without mentioning the other four buildings The client says that the roof space is always in use by one group or another but more so during the summer months, when they enjoy having impromptu 'get together' that could last well into the evening with a barbeque, It is easy to see why it is described as a nice space for just hanging out after work.

higher then the west end for entertainment. It is a good place to live work and play.

버비지 하우스

버비지 하우스는 올드 스트리트와 그레이트 웨스턴 스트리트 사이에 자리잡고 있으며, 엎어지면 코 닿을 거리에 비숍스 게이트, 무어게이트, 그리고 리버풀 스트리트 스테이션이 위치하고 있다.

버비지 하우스 주변에는 모두 4동의 건물이 버비지 하우스보다 살짝 낮은 높이로 자리하고 있다. 이들 가운데 둘은 버비지 하우스의 양 옆에, 나머지 둘은 뒷편에 위치하는데, 이들 두 건물의 출입구는 샬롯 스트리트와 연결된다. 옥상을 통해 다른 건물로 넘어가 비상구를 이용할 수 있도록 건물들이 연결되어 있다.

현재 버비지 하우스에 입주한 업체들은 매우 다양하다. 금융업을 하는 액션 비자스, 음악에 특화된 보나 피데 스튜디오, 다이버스 인테리어 아키텍트, 그리고 자신들이 "새로운 제품의 디자인과 개발"을 하고 있다고 설명하는 럭키 바이트를 비롯 많은 업체들이 여기 입주하고 있다. 나머지 네 개의 건물에 대한 언급은 피하도록 하자.

의뢰인의 얘기에 따르면 옥상 공간은 언제나 누군가에 의해 사용되고 있으며, 특히 여름이 되면 그 이용률이 더 높아진다고 하는데, 함께 즐거운 오후를 만끽하다가 자연스럽게 바비큐 파티를 벌이기도 한다는 것이다. 이곳이 왜 멋진 장소인지 이해하는 것은 결코 어려운 일이 아니다.

웨스트 엔드보다도 즐길거리가 더 많은 이 지역은 주거, 업무, 그리고 유흥에 있어서 충분한 경쟁력을 갖추고 있다.

Zillah Akawu-Irmiya
zillah7_4@hotmail.com

Shadow

The rooftop is located in very diverse and creative area. The Burgabe House building alone contains artists from different fields of art working in different design businesses. We can find there companies that specialize in: Interior Design, Architecture, Web Design, Fashion Design, Film making and Music making.

옥상은 매우 다채롭고 창의성이 넘치는 지역에 자리잡고 있다. 버비지 하우스만 하더라도 다양한 디자인 업계에서 일하는 다양한 성격의 아티스트들이 입주해 있다. 버비지 하우스는 인테리어 디자인, 건축, 웹디자인, 패션디자인, 영화제작, 그리고 음반제작 등 특화된 분야가 각기 다른 다양한 업체들을 만나볼 수 있는 곳이기도 하다.

Around the building there. are different Galleries, Estate Agents, Architecture and Interior Design companies, a number of cafés and other meeting places, and Fashion College.
The area is also reach in different graffiti works.

건물 주변에는 갤러리, 부동산 중계소, 건축 및 인테리어 디자인 업체, 수많은 까페들, 그리고 패션 학교가 들어서 있다. 또한 이 지역에서는 다채로운 그래피티 작품을 만나볼 수 있다.

View showing Curtain Road in the Shoreditch area.

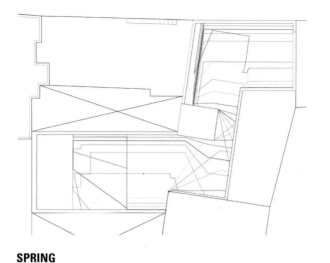

SPRING

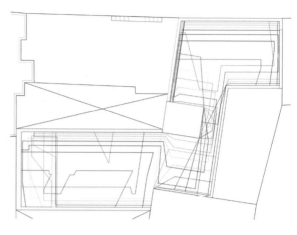

SUMMER

The aim was to create a space for artists from different fields of art, but sharing the same passion.

People entering the rooftop will be distracted and encouraged to interpret the place, play with it and share different points of views – just like a child entering the playground.

These plans based on the seasons, Spring, Summer, Autumn and Winter show the lines of light and shadow that delineate the space within

프로젝트의 목표는 각기 다른 예술 분야에서 활동하지만, 공통의 열정을 공유하는 아티스트들을 위한 공간을 창조하는 것이다.
옥상 공간에 들어선 사람들은, 마치 놀이터를 찾은 어린이처럼, 공간을 해석하고, 이를 즐기며, 다양한 관점을 함께 나눌 수 있는 기회를 갖게 될 것이다.

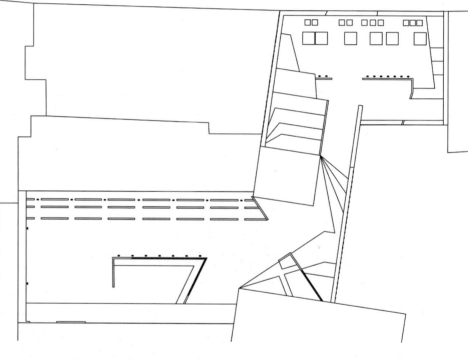

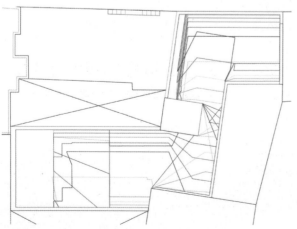

AUTUMN

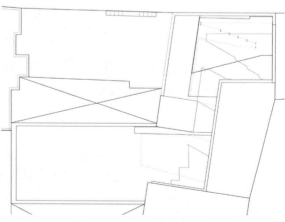

WINTER

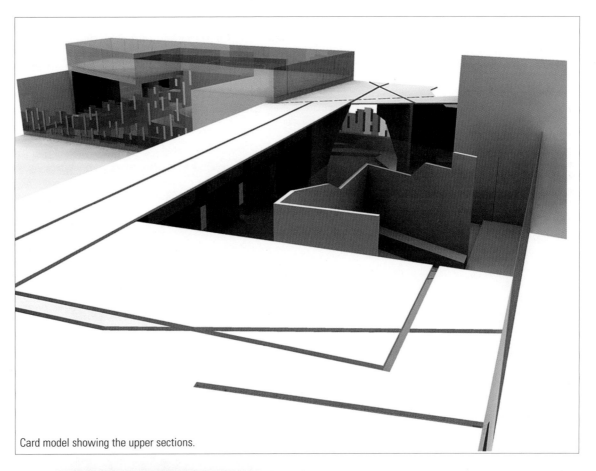

Card model showing the upper sections.

Card model showing space within.

Materials

The materials that are used on the rooftop are: existing already bricks, concrete, glass and corian. I wanted to show the contrast between the rough and smooth character of materials, to make the area more interesting.

By its structure it distracts people from their work giving them time for their selves, to recover.

As the leading word is Shadow, I decided to observe the shape and the movement of the shadow on the rooftop, every hour during days of year's solstice, that is on: 21st of March, 21st of June, 23rd of September, and 21st of December.

By this method interesting and dynamic combinations of lines were created. Then came the selection of the most interesting lines that do cooperate with each other. After that all what had to be done was to create the roof, walls, seating places, lighting and walking spaces and put them in the place of the lines

자재

옥상에 사용되는 자재는 기사용중인 벽돌, 콘트리트, 유리와 코리안 등이다. 나는 부드러운 소재와 거친 소재의 대비를 보여주고 싶다. 이를 통해 공간은 모두 흥미진진하게 변할 것이다.

그 구조를 통해 공간은 사람들로 하여금 자신이 하던 일을 잠시 있고, 자신만의 시간을 보냄으로써 자아를 회복할 수 있는 시간을 제공해준다.

주제인 그림자에서 눈치채 수 있듯이 나는 옥상 위의 그림자가 매년 춘분, 하지, 추분, 동지인 날에 보여주는 모양과 움직임을 관찰하기로 했다.

이 방법을 통해 흥미롭고 역동적인 선들의 조합을 만들어낼 수 있었다. 그 뒤 서로 완벽한 조화를 이루어내는 가장 흥미로운 선들을 골라낼 수 있었다. 이후 해야했던 일은 지붕, 벽, 앉을 자리, 조명, 그리고 걸어다닐 수 있는 공간을 만들고, 그것들을 선에 맞춰 배치하는 것이다.

Plan showing the partial Roof structure.

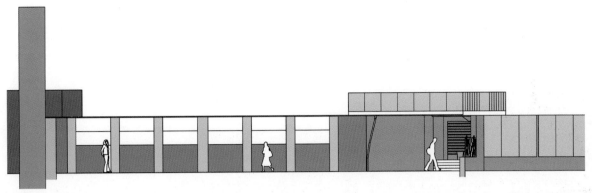

Elevation :Rear View

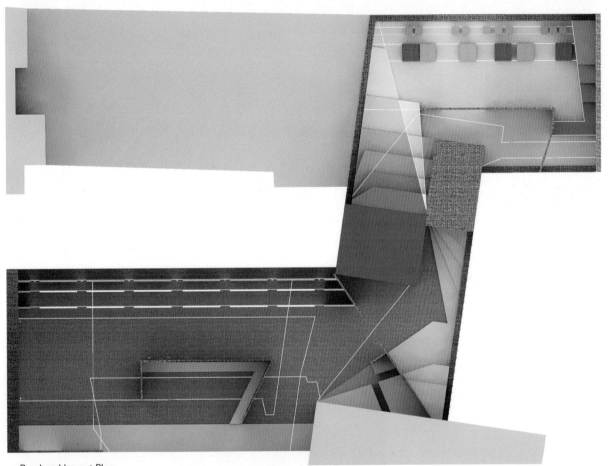

Rendered Layout Plan

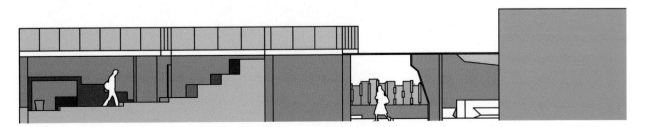

Elevation: Front View

The People

The roof top is mostly used by artists, so it should be designed for them, which means that it should reflect Art and relate to it.

It should also work as a link to them and between them, gathering people together and disconnecting them with their work.

This gives you an idea of a creative place that needs to suit everybody or at least most of the people using the area. A place away from work during break-time. A place to talk, relax and recover. A place that is in the city but above it, closer to the sky. A flexible space not only for break-times, but also for different meetings and parties or as a showroom for exhibitions.

사람들

옥상은 주로 아티스트들에 의해 사용되고 있다. 따라서 옥상 디자인도 이들을 배려해 이뤄져야 하는데, 즉 예술을 반영하고, 이와 연관되어야 한다.

더불어 옥상은 이들을 이어주는 가교로서의 역할을 하는 동시에 업무로부터 잠시 벗어날 수 있는 기회를 제공해야 한다. 이를 통해 공간을 사용하는 모든 사람들, 혹은 적어도 대부분의 사람들을 만족시킬 수 있는 공간의 모습이 어떠해야할 지 상상할 수 있다. 휴식 시간에 업무에서 벗어나 이야기를 나누고, 휴식을 취하고, 기운을 회복할 수 있는 공간이 바로 그것이다. 도심 속에 자리 잡고 있지만, 위치가 높아 하늘에 가까이 닿은 곳. 휴식 시간을 보내는 것 뿐만 아니라 다양한 미팅이나 파티, 혹은 전시 공간으로 활용할 수 있는 융통성이 존재하는 공간!

The main word for my concept is: Shadow. Shadow gives a shelter and protects from the burning sun, giving a moment for relaxation. The rooftop works in similar way, but it gives a time out from work, because every work can be tiring, even the one that you really like.

나의 디자인 컨셉트를 한 마디로 표현하자면 그림자가 될 것이다. 그림자는 우리에게 휴식처를 제공하고, 강렬한 태양으로부터 보호해주며, 휴식을 위한 시간을 마련해준다. 옥상도 이와 마찬가지이지만, 이와 더불어 업무로부터 벗어날 수 있는 시간을 가져다준다. 모든 일은 힘들기 마련이다. 심지어 당신이 진정 좋아하는 일일지라도 말이다.

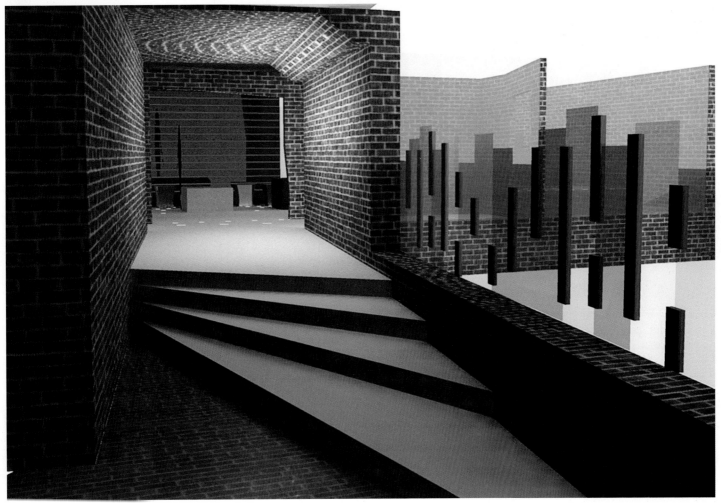

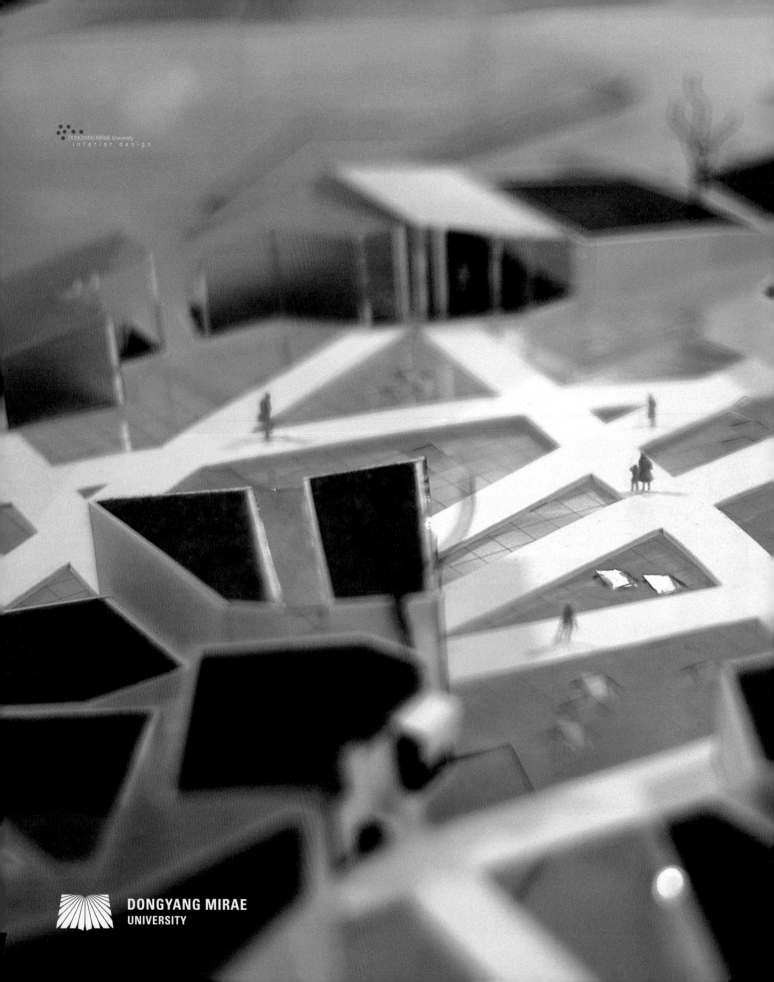

DONGYANG MIRAE UNIVERSITY
Department of Interior Design

2010 sophomore design studio

The 2nd Hidden Space Project
URBAN ROOFTOPS

동양 미래대학 실내디자인과
2010 2학년 2학기 설계스튜디오 결과물

152-714
DONGYANG MIRAE University Dept.of Interior Design
62-160, gocheok-dong, guro-gu, seou, korea
phone + 82 2 2610 1962
fax + 82 2 2610 1960

http://interior.dongyang.ac.kr

학부장 **김홍기 교수**
동양미래대학 디자인학부 실내디자인과
hkkim@dongyang.ac.kr

서 문

"숨겨진 공간, 옥상"이라는 주제로 런던 메트로폴리탄 대학과 동양미래대학이
작품집을 발간하고 교류전을 갖게 된 것은 매우 의미있는 일이 아닐 수 없다.
서로 다른 문화와 역사를 갖고 있지만 시공을 초월하여
'옥상'이라는 주제로 머리를 맞댄 것이다.
근대 이전까지 지붕은 지역성과 민족성을 상징하는 요소였다.
두모성당의 돔에서 내려다 보는 피렌체는 아름답기 그지없고,
프라하 성에서 내려다보는 도시의 모습은 보석과도 같다.
그러나 근대화 이후, 철근콘크리트 문화가 착종되면서
건축의 형식은 바뀌고 지붕은 새로운 삶터의 공간으로 재해석된다.
르 꼬르뷔지에는 '도미노 구조'를 내세워 옥상정원의 이론을 내세웠고,
마르세이유 아파트, 유니테 다비타시옹에서 옥상은
아이들의 놀이시설과 수영장으로 진화한다.
서민들의 애환이 담긴 달동네 옥탑방에서
고급주택을 일컫는 초고층 빌딩의 펜트하우스에 이르기까지
옥상은 다양화 진화과정을 거친다.
과연 숨겨진 공간, 옥상은 어디까지 발전할 수 있을까?
이번 전시는 그러한 욕망을 드러내는 작업의 결과물이다.
다소는 실험적이고 현실을 넘어서는 제안이 있는 것은
창조를 향한 꿈이 있기 때문이다.
2010년도에 이어 두 번째로 기획된 한국과 영국의 국가교류전,
런던 전시회를 축하하며
참여한 학생과 튜터들에게 환희의 박수를 보낸다.

Head of Interior Design Department | **Prof. Kim, Hong Ki**
Department of Interior Design DONGYANG MIRAE University
hkkim@dongyang.ac.kr

Preface

It was a task of great significance that London Metropolitan University and Dongyang Mirae University published a collection of works and had an exchange exhibition with the topic of "A Hidden Space, a Rooftop".
They put their heads together with the topic of 'A Rooftop' beyond time and space even if they had different cultures and histories.
A rooftop up to pre-modern was an element which symbolized regionality and ethnicity. Firenze viewed from a dome of Duomo Chapel is beautiful beyond all description.
A picture of a city viewed from Prague Castle is a real gem.
However, the form of architecture was changed and a rooftop was reinterpreted as a space of a new livelihood as ferroconcrete culture was entangled after modernization.
As Le Corbusier proposed a theory of a roof garden by advocating 'Domino Structure', a rooftop was evolved into playing facilities for children and swimming pool in Marseille Apartment called Unitè d'Habitation.
A rooftop had gone through an evolutionary process of diversifying from a rooftop house in a poor hillside village which embraces the joys and sorrow of ordinary people to a penthouse of a high-rise building called an expensive house.
Indeed, how far can a hidden space, a rooftop, be developed?
This exhibit is the result of works which show such desire. There is a proposal which is somewhat experimental and beyond reality because there is a dream which goes off to creation.
I give applause of joy to students and tutors who participated in by celebrating London exhibition, a national exchange exhibition between Korea and England planed for the second time following 2010.

잡을 수 없는 불꽃같은 공간을 위하여..
For an Uncatchable Flame-like Space..

2학년 설계담당 **박영태 교수**
동양미래대학 디자인학부 실내디자인과

responsible for sophomores | Prof. Park , Young Tae
Department of Interior Design DONGYANG MIRAE University
robotomy@naver.com

Hidden Space project - Episode 2 "Urban Rooftops"

숨겨진 공간 프로젝트
– 에피소드 2 "도심 옥상공간"

Hidden Space Project는 2010년 7월 런던 아키텍 페스티벌에 참여했던 국제교류전인 "Living Bridges"로 시작되었다.
우리주변에 숨겨진 공간으로서의 "삶을 같이 영유하는 다리" 라는 공간을 주제로 접근한 첫번째 프로젝트에 이어 두번째로 진행되는 프로젝트는 "URBAN ROOF-TOPS(도심지 옥상)" 공간이 그 대상이다.

2010년도 2학년 2학기에 2개의 스튜디오로 진행된 내용으로, 1주에 6시간, 전체 15주 100일 동안 총 23개의 작품이 진행되었다.
발표와 토론, 2차원 표현기법(스케치와 도면, 컴퓨터그래픽)과 3차원 표현기법(3D 컴퓨터 모델링, 렌더링)과 모형제작과정을 통해 전체수업을 마무리하였다. 이 스튜디오는 작품의 개념과 함께 형식(표현)에 대한 훈련 역시 매우 큰 의미가 있는 종합적인 성격의 스튜디오 수업이다.

약 3주간의 현장조사와 거주자, 사용자들과의 인터뷰 그리고 2차에 걸친 내외부 크리틱을 통해 프로세스 자체에 많은 의미가 부여될 수 있도록 진행하였다.
특히, 전년도 Living Bridge와 마찬가지로 이 주제는 사회와 문화, 커뮤니티를 핵심으로 접근하였다.

최근 도심공간의 활용과 에너지와 생태 등이 이슈가 되면서 조경공간이나, 그 밖에 휴게공간, 혹은 좀 더 공격적인 상업공간과 또 작물 등의 재배공간으로서까지 기능적 확장이 대두되고 있는 옥상공간에 대한 관심은 시기적으로 그리 대단한 이슈라고 볼 수는 없었지만, 이 프로젝트의 목적은 모더니즘의 습관적 관성으로부터 좀 더 거리를 두고 옥상공간의 본질을 찾아보자는 데에 있었다. 많은 부작용이 나타나기는 했으나 시작단계의 화두는 근본적인 옥상에 대한 선입견과 규범들을 넘어서는 데에 있었다. 이를 통해 기존의 고질적인 옥상공간들의 동일화 양상을 벗어나기 위해 모두 다 노력하였다.

Hidden Space project
– Episode 2 "Urban Rooftops"

The Hidden Space project was started with an international exchange exhibition, "Living Bridges", participated in London Architeck festival on July, 2010. It was a project of "Urban Rooftops" secondly carried on following the first project with a topic of a space called "Bridges" as a space hidden around us.

A total of 23 works were showed for 6 hours a week, a total of 15 weeks, 100days as contents done in 2 studios at the second semester of sophomore year in 2010. The whole class was finished through presentations and discussions, two-dimensional expressive techniques (sketch, floor plan, and computer graphic), three-dimensional expressive techniques (3D computer modeling and rendering), and the process of modeling. The training about the form (expression) and the concept of works was also an important studio course with comprehensive characteristic.

It was gone on to give many meanings to the process itself through an about three-week field study, interviews with residents and users, and internal and external criticism done twice. The subject was approached with society, culture, and community as key words like "Living Bridge" of the previous year.

It was not a great issue in time, the interest in a rooftop space in which a function expansion has become to the fore up to a landscape space, a resting space, a more aggressive commercial space, or a space for crop cultivation as the application of urban space, energy, and ecology had become issues. However, the purpose of this project was to find an essence of a rooftop space by keeping more distance from the habitual inertia of modernism. The inchoate topic was to get over prejudice and standard about a fundamental rooftop even if there were many by-products. Everybody made efforts to get out of existing chronic standardized patterns of rooftop spaces.

옥상공간에 대한 사회, 문화적 접근

우선 풍족하고, 여유로운 공간으로서의 옥상의 활용보다는 사회적 약자들의 공간적 결핍과 제약을 위해 접근을 시도하였다.

쪽방촌, 고시촌, 외국인 노동자 주거지 등의 고밀도의 척박한 공간으로서 어떤 새로운 시도가 불가한 그러한 공간에서 옥상이라는 공간을 통해 변화될 수 있는 사회적 약자들의 생기를 상상했으며, 오래된 고밀도 오피스들의 옥상, 근린 생활시설들의 밀집된 낙후된 상가건물들의 옥상, 재건축이나 재개발이 불가한 오래된 저층 아파트 단지들의 옥상, 일정 지역에서 그 지역의 상징성을 유지하며 역사적 가치를 지닌 건물(폐학교, 폐공장등으로)로서 옥상공간의 접근으로 재활성화의 가능성을 모색하였다.

또 개별 건물들의 옥상 자체가 될 수도 있으며, 혹은 옥상과 옥상사이가 될 수도 있을 것이며, 상가 건물의 옥상과 주거공간의 옥상의 사이에서의 새로운 상상 등을 단서로 하였다.

단, 우리가 집중한 점은 위에 열거한 공간적 사회적 약자들에 대한 이 시대안에서의 적합한 정의를 내린 후 행복한 사용자로서의 그들의 지위를 재 위치하는 것이었다.

A Socio-cultural Approach to a Rooftop Space

An Approach was first attempted for the special shortage and restriction of the disadvantaged rather than for the application of a rooftop as an abundant and leisurely space. We imagined the vitality of the disadvantage, which could be changed through a space called a rooftop in as a barren space with high density where any new attempt was impossible such as a dosshouse, a village for examination preparation, and a residence for foreign laborers. We attempted to seek the possibility of revitalization by access to a rooftop as a building (a ruined school and factory) which kept the symbolism of that region and historical value in rooftops of old office buildings with high density, rooftops of crowded and underdeveloped commercial buildings in neighborhood living facilities, and rooftops, certain areas of old low-rising apartment complexes where reconstruction or redevelopment was impossible.

The Clue was new imagination on rooftops of individual buildings, a space between rooftops, or a space between a rooftop of a commercial building and a rooftop of a residential space. Our main focused point was that their positions as happy users were re-located after giving an appropriate definition about the spatial disadvantaged enumerated above in this era.

잡히지 않는 불꽃같은 공간을 위하여

수업과정을 통해 하늘을 향해 한없이 열린공간으로서의 옥상이라는 공간이 가지는 가능성에 대해서 많은 한계를 가지고 있음을 확인 할 수 있었다. 다른 공간에 비해 옥상공간 상상력의 현실화는 법규와 구조 등의 문제이전에 보다 기본적인 제약이 많음을 거듭 확인하였다. 최상부에 존재하는 공간이 가지는 접근성이라는 조건과 이로 인한 구축적 현실성 등이 큰 고민거리로 작용하였으며, 사회적 약자인 사용자들의 생활체계속으로 흡수될 수 있는 공간으로서의 기능과 프로그램 역시 작고 소박하지만 실제적인 작동과 유지가 우선해야 함이 그리 쉬운 문제는 아니었다.

옥상공간을 강화할 수 있는 많은 개념들과 상상력이 동원되었으며, 그 공간의 구체화에 있어서는 적지않은 진통 역시 동반되었다. 특히 구축에 있어서 학생들에게 옥상이라는 공간이 구체적으로 대상화되는 과정에서는 진행된 모든 개념적 가능성의 가치가 무의미해지기도 했으며, 또 기존에 자리했던, 수많은 미디어로부터 우러나온 여러 유형의 고질적 잔상들의 개입으로 프로젝트의 본질을 왜곡할 수도 있는 부분이었다.

다중(多衆)이 사용하는 다양체로서의 옥상공간. 형상으로서 잘 다듬어지기보다는, 시간의 흐름과 함께 배어나와야 할 사용자들의 요구와 그런 요구로부터 재활성화될 수 있는 유기체로서의 분위기를 내재한 질료적 공간이 되도록 의식하였다.

    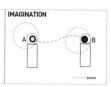

15주간의 짧고도 긴 시간동안 모든 구성원들이 이 잡힐 듯 말듯 한 한 순간의 불꽃의 공간, 결코 정지되어서는 안 될 생성의 공간을 위해 가슴 깊이 몰입하였다. 돌이켜보면 옥상공간은 공간의 성격상 첫번째 주제인 Living Bridges에 비해서 많이 어려웠던 것 같다.

끝까지 스튜디오를 함께 운영해 준 김석영 교수와 곁에서 후원해 준 실내디자인 학부장님과 동료교수님들에게도 감사드린다.

또, HIDDEN SPACE라는 주제로 작품을 공유하며 비록 거리상의 문제로 유기적 교감이 아쉽기는 했으나, 시작과 마무리를 함께 해 준 런던 메트로폴리탄 대학의 교수와 학생들에게도 감사드리며, 무엇보다도 15주동안 열정과 지혜로 함께한 실내디자인과 2학년 구성원 모두에게 이 작품집의 의미가 깊이 새겨졌으면 바램이다.

is, a substance space which contained demands of users which should be come out with the passage of time and an atmosphere of an organism activatable from demands of users. All the members had been absorbed in works to create a space of the flame of the moment difficult to catch, a space of the creation which should never be stopped for a 15 week- short, but long time. Thinking back on it now, the "Rooftop Space" project was remembered to have much more pains than "Living Bridges", the first topic of "Hidden Space"

I want to express my heartfelt gratitude to Professor Sek Young Kim who had managed a studio together to the end, the dean of Interior Design Department and fellow professors who support me by my side. I

03 문학작가로서 새로운 옥상공간에 대한 시나리오 작성하기
기존의 옥상에서 생각되어질 수 있는 것들에 대한 정리

· 연결/소통/교감/커뮤니케이션...
  – 물리적 거리, 문화, 지역정체성과 정서..
· 교통/전망/휴식...
  – 차량과 보행자, 전망대, 지역(도시)경관감상
· 이동운동/연속/흐름/속도/시퀀스...
  – 소프트웨어의 특성의 개념적·체계적 접근
· 독립/부유/개방/노출/취약...
  – 하드웨어의 특성의 개념적, 체계적 접근

10페이지 분량으로 다음과 같은 내용을 조별로 발표한다.
01 접근 배경과 조건
02 SITE(도시,단지)의 특징 종합분석
02 해당공간의 사회적, 공간적 약자들에 대한 종합적 분석
   개별인터뷰 및 FGI 진행

   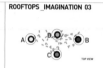 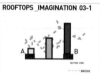 

For an Uncatchable Flame-like Space

It was ascertainable that there was many limitations on the availability of a space called a rooftop as a space infinitely open toward the sky through this course. It was identified once again that the actualization of imagination over a rooftop space had more fundamental restrictions before legal and structural problems than other spaces. Big questions were a condition called accessibility of a space which existed at the super-top and the accompanying constructional actualization. It was also not an easy problem to give priority to its actual operation and maintenance even if a function as a space absorbed into the living system of users, the advantaged, and program were very small and humble.

Many concepts and imagination needed to strengthen a rooftop space were mobilized and no little pains were accompanied in giving shape to that space. Especially, in construction, the values of all the conceptual possibilities processed became meaningless to students in the process that a space called a rooftop was concretely objectified. It was a part that the essence of this project could be distorted by the interference of many types of existing chronic afterimages derived from lots of media.

It was considered to make a rooftop space as a manifold space used by a large number of people, that

also gave thanks to professors and students of London Metropolitan University who joined from start to finish by sharing works with the topic of "Hidden Space" even if I felt the lack of close relationship due to a regional distance problem.

Above all things, I hope that all the sophomores at Interior Design Department keep the meaning of this collection in their minds.

스튜디오 진행내용

주요 수업진행 내용과 과제내용은 다음과 같다.

핵심어

SOCIAL ISSUE
COMMUNITY BASE
SUSTAINABLE

작은 옥상, 큰 옥상, 한 개의 옥상, 여러 개의 옥상, 낡은 옥상.....그리고, 동시대 우리들의 생활체계에 대해서 생각해보기.
우리의 옥상
01 생산(자)/소비(자)/거주(자)/사용(자)/관찰(자)/비판(자) 등 각각의 입장에서 정의내려보기
02 옥상과 생활속의 콘텐츠에 대해 생각해보기
   의/식/주.....
   먹고, 자고, 마시고, 놀고, 쉬고, 감상하고, 배설하고,
   사랑하고, 사고, 팔고, 춤추고, 노래하고, 운동하고,
   산책하고, 공부하고, 대화(토론)하고.......

03 옥상이 위치한 공간(건물)에 대한 분석
04 해당 공간의 옥상공간(들)으로서의 특징과 이슈
05 옥상공간의 용도
06 옥상 공간전개에 대한 컨셉(SWOT제시)

스튜디오 결과물 및 평가
00 PROJECT RECORDING(15주간의 기록)
작업내용 DOCUMENTATION – 주별로 웹하드에 업로드 작업과정과 자료조사, 답사 그 밖에 작업전반에 대한 글과 그림을 폴더별로 정리. 작업자가 포함 된 작업전반에 대한 사진촬영 필수

02 중간평가
판넬 – 700X2100 세로 족자형.
모형 – 900X900크기에 담길 수 있는 스케일 모형

03 기말 평가
좋은 생각을 전달할 수 있는 표현물
판넬 – A0 세로 족자형.
모형 – 900X900크기에 담길 수 있는 스케일 모형
영상 – 3인이상의 경우 경우 동영상 표현 필수

04 최종 작품집용 결과물 제출
국문, 영문작성
작품당 6~10PAGE 작성
PHOTOSHOP PSD 화일과 / INDESIGN 화일로 제출
프로젝트레코딩 내용물 제출

Module Summary (2010.09.07)

00 REQUIRED COMPULSORY SUBJECTS
Sophomore "Inteiror Design Studio"

4 Of 18 Credit Modules in a Semester
· 6 Hours (tuesday) / 1 Week
· Team Project - 2 Students / 1 Team
· Total 23 Projects

01 THEME
"URBAN ROOFTOPS"

02 PROCESS & OUTPUT
· Research & Analysis (include FGI & survey, SWOT)
· Space Scenario making (like a cartoon presentation)
· Sketch(analog & digital method) process
· Complex Space Program(zoning, planning....)
· Drawing
· 3D Simulation
· Scale Modeling
· Panel Presentation
& Every Process(Photo & Text) Recording - for the Final Documentation

03 SCHEDULE
· 2010.9. 1st Week - Kick off & Orientation
· 2010.10.Last Week - Mid Term Critic (Small Scale Exhibition & Panel presentation)
· 2010 12.2nd Week - Final Exhibition

04 KEY WORD
· SOCIAL DESIGN
· COMMUNITY BASE
· SUSTAINABLE
· PROSTHESIS
· PERFORMANCE
· PROJECTIVE & REFLEXIVE
· RESONANCE
· SITE-SPECIFICITY

전임강사 **김석영**
동양미래대학 디지인학부 실내디자인과

Full Time Lecture Kim, Suk Young
Department of Interior Design DONGYANG MIRAE University
afour0425@dongyang.ac.kr

앨리스적 공간을 위하여…
For an Alice Style Space.

욕하면서 닮는다.

평소 마음에 들지 않았던 상대방의 행동을 나 자신에게서 발견하거나, 혹은 꼴도 보기 싫었던 그 사람에게서 나의 모습을 발견할 때가 있다. 곁눈질로 상대를 힐끔 돌아보며 뭐라 설명하기 힘든 미묘한 감정을 느낀다. 불편한 동질감. 내가 그렇게도 불평을 늘어 놓았던 그 녀석과 어느덧 닮아버린 나…… 그건 나의 의지일까?

서로 스치고 지나가는 일상에서의 우리의 말과 행동은 좋든 싫든 간에 주어진 환경의 그물망과 관계한다. 그리고 그 안에서 서로에게 영향을 주고 받는다. 그 결과가 지금의 내 모습니다. 모든 관계들로부터 자유로울 수 있다는 것은 오직 세상을 만든 신에게만 주어지는 특권이 아닐까? 실내디자인이라는 단편적인 학문 분야에서 선생과 제자 간에 형성되는 관계는 일상적 관계가 아닌 학문적 목적을 위한 특수한 이해 관계이다. 따라서 디자인 스튜디오 수업에서의 방향 설정은 선생과 제자가 각자의 시각으로 현실적 공간을 다양한 창의적 시도로 발전시킬 수 있는 가능성을 제시하는 과정의 시작이다.

Resemble by swearing world.

There are many cases that behaviors of the other party who is not to my taste are found in me or my outlook can be discovered in that person whose very sight makes me quite disgusting. I feel the odd mix of emotions beyond explanations by looking at that person out of the corner of my eyes. An unease sense of kinship. Myself who became looked like him whom I kept making complaints that way without my knowledge. Was it my will? Our words and behaviors in our daily lives brushed through are related to the mesh of the given environment for better or worse. We influence each other in that. The result is the way I look now. What can be free from all the relations is the privilege given only to God who created the world, isn't it? The relationship formed between a teacher and students is special interests for academic purposes, not a causal one in the fragmentary academic field, that is, interior design. Therefore, the set-up of direction in a design studio class is the start of the process which suggests the possibility that a teacher and students

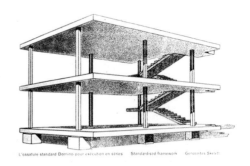

L'ossature standard Domino pour exécution en séries   Standardised framework   Geonontres Sketch

부여하는 중요한 역할을 해왔다. 서양 건축의 전범인 파르테논도 강직하게 늘어선 기둥 위에 중심을 잡고 올려져 있는 패디먼트의 형태가 두드러진 특성으로 우리에게 각인되어 있다. 지붕은 상징적인 의미로서 강조될 뿐 그 위의 공간은 그저 허공에 지나지 않았다.

그러나 근대에 이르러 건축의 구조와 재료가 발달하면서 삼각형의 단면이 가져다 주는 안정감을 극복하게 되었다. 수평면의 지붕, 즉 옥상이 등장하게 된 것이다. 목적성을 부여 받은 건축공간의 영역으로부터 제외되어 왔던 '지붕' 위의 허공이 이제는 인간의 활동을 수용하는 적극적 의미의 '옥상' 으로 변화하게 된 것이다.

뒤늦게 생활 속에 등장한 공간이면서 외부로 노출된 공간이어서 일까? 옥상은 그 누구에게도 속하지 않는 특성을 갖고 있다. 따라서 일상에서 소외된 빈 공간을 어떻게 사람들이 공유하고 친근한 관계망을 형성하는 장소로 변화시킬 수 있는지에 대한 제안이 필요했다. 어느 누구의 것도 아니기에 누구나 사용할 수 있는 공간을 계획하고자 했다. 내가 아닌 모두를 위한 우리의 공간으로 프로젝트의 방향을 설정했다.

develop a realistic space with creative attempts from their own viewpoints.

모두를 위한 하나의 공간

'옥상의 주인은 누구인가' 라는 질문에서 프로젝트의 방향을 구상하기 시작했다. 첫 단계에서, 디자인의 물리적 대상은 명료했지만 인적 대상은 모호했다. 디자인의 구체적인 전개에 앞서 우리에게 옥상이란 어떤 공간인지 되짚어 보았다.

선사시대에서부터 19세기에 이르기까지 지붕은 적극적인 사용목적을 갖는 단일 공간이기보다는 외부와 내부의 구분을 목적으로 설치된 건축요소였다. 물론 중앙부가 솟아오른 박공이나 돔의 형태는 건축물에 상징적 의미를

A Single Space for Everyone

The direction of this project was started to plan from a question of "Who is an owner of a rooftop?" At the first stage, the physical object of a design was clear, but the human object was vague. It was reviewed what kind of a space a rooftop is to us prior to the concrete display of a design.

A rooftop had been an architectural factor installed for a purpose of dividing the inside and outside rather than a single space with the purpose of its active use from the prehistoric times to the 19th century. The form of a gable or dome with the soaring central part played an important role of giving the symbolic meaning to the structure. The Parthenon, a typical model of Western architecture, is vividly engraved on our mind through characteristic which a form of pediment kept its balance and placed on pillars uprightly lined up is remarkable.

However, the stability brought by the side of a triangle was overcome as the structure and materials of architecture have been developed in modern times. A roof of a horizontal plane, that is, a rooftop came to appear. An inane space on a 'roof' excluded from areas of the architectural space, which the purpose of use was given becomes now changed as a 'roof' with active meaning, which contained human activities. Is it because it is a space which appeared in human life too late and a space exposed to the outside? A roof had its own unique characteristic which does not belong to anything. There is a need of some proposal to suggest how an empty space excluded from everyday life can be changed into a place where people can share and form close networks. It was attempted to plan a space where anybody can use because it does not belong to anybody. The direction of the project was set-up to create our space for everyone, not for me.

다수의 주체에 의해 끊임없이 변화하는 공간

옥상에 정원을 두어, 일반 도로로부터 분리시켜, 생활 공간으로 사용한 건축가가 근대의 거장 르 꼬르뷔제였다. 그는 자신의 건축작업에 필요한 몇 가지 원칙을 선포하면서 옥상공간의 사용을 포함시켰다. 그는 옥상공간을 적극적인 생활공간으로 계획하기는 했지만, 한 개인을 위한 곳이 아닌 공공성을 지닌 공간으로 실현시켰다. 빌라 사보아에서는 건축적 산책로인 경사로의 끝자락에 위치한 정원으로 계획했으며, 마르세이유의 위니떼 다비따시옹에서는 어린이들을 위한 보육시설과 물놀이 공간을 배치했다.

해 놓았던 것과 같이 말이다. 이것은 무수히 많은 가능성을 모두 배제한 채 오로지 하나의 해법만을 제시하는 단답형 공간이다. 그렇다면 오늘 우리는 생활 속에 새롭게 등장한 옥상이란 공간에서 이러한 한계를 어떻게 극복할 수 있을까? 한 사람의 아이디어가 아닌 다수의 의견이 투영되는 공간을 어떻게 구현할 수 있을까?

ferent idea completely free from the early design concept of a rooftop proposed by Le Corbusier even if a century went by. Therefore, it was attempted to seek the result of a rooftop design which the design thought of today was reflected a new time was required. The master of modern architecture was a planner who created the 'Machine to Live' which provided the universal residence to anybody anywhere through all ages. An architect proposed the most appropriate functional space by prospecting and accepting demands and behaviors of users. He was a man of undoubted ability who stood at an observer's position, viewing the whole picture at a

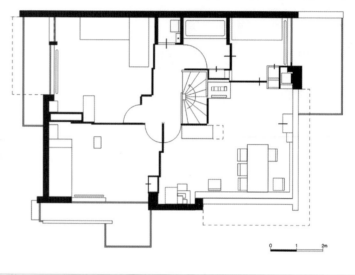

20세기 초 이미 근대건축의 대가는 옥상을 새롭게 인식했고, 중요한 디자인 요소로 지목했으며 여러 가지 실험적 실천을 감행했다. 한 세기가 흘렀지만, 르 꼬르뷔제가 제안한 초창기 옥상 디자인 개념으로부터 완전히 벗어나 다른 아이디어를 제안한다는 것은 간단한 문제가 아니다. 따라서 오늘날의 디자인 사고가 반영된, 새로운 시점을 요구하는 옥상 디자인의 결과를 모색해야만 했다. 근대의 건축가는 어느 시대, 어느 장소 그리고 어느 누구에게나 보편적 주거를 제공하는 '살기 위한 기계'를 만드는 계획자였다. 건축가는 사용자의 요구와 행위를 예견하고 수용해 가장 적절한 기능적 공간을 제안했다. 건축가는 원근법에서 – 그림 전체를 한 눈에 조망하는 – 관찰자 위치에 서 있는 능력의 소유자였다. 소수의 거주자를 대상으로 하는 프로젝트에서 그들의 요구와 행위를 반영해 공간을 제안하는 것은 가능하다. 그러나 도시가 발달하면서 수백 명 혹은 수천 명이 함께 사용하는 대규모 프로젝트가 출현하게 되었고 사용자의 의지와 욕구도 다양하게 확대되었다. 따라서 건축가는 합리성과 권위를 이용해 사람들의 생활을 재단하고 조정하는 역할을 하게 된 것이다. 마치 영화 메트릭스 속 아키텍쳐가 무소불위의 능력으로 완벽해 보이는 세상을 프로그래밍

A Endlessly Changing Space by Many Subjects

Le Corbusier, a master of modern architecture, was an architect who utilized a rooftop as a living space by separating it from a public road as he put a garden on a rooftop. He included the use of a rooftop by announcing several principles needed for his own building operations. He realized a rooftop as a space with the publicness, not one for an individual even if he planned a rooftop as an active living space. He planned a garden located at the edge of a slope way, an architectural promenade, in Villa Savoye and arranged a child-care facicilty and a water play space for children in Unite d`Habitation of Marseille.

The master of modern architecture had already recognized a rooftop newly in the early 20th century, pointed out a rooftop as one of important design factors, and carried out several experimental practices. It was not a simple problem to propose a dif-

glance, in perspective.

It is possible to propose a space by reflecting demands and behaviors of a few residents for the project for them. The large project for hundreds or thousands of people has appeared, and the will and desires of users have diversely been expanded as the city has been developed. Therefore, the architect becomes to play a role in designing and adjusting people's life by using his or her rationality and authority.

It was as if an architect programmed the world perfect with absolute ability in a movie, 'Matrix'. This is a short-answer question type space which suggests only a single answer by excluding all the multitudinous possibilities. If so, how can we overcome this limit in a space called a rooftop which newly appeared in our lives today? How can we realize a space in which the views of the majority, not only an idea of one person, are projected?

다수의 주인이 함께 공유하는 옥상은 다양한 욕구가 담길 수 있어야 한다. 그러기 위해서는 주어진 공간에 무엇인가를 덧붙이는 것이 아닌, 가능성을 위해 비우는 작업이 우선되어야 한다. 그러나 디자인의 결과로 극단적 공허를 제안하는 것은 무의미한 공회전 만을 낳게 된다. 따라서 계획을 담당한 건축가나 디자이너는 주어진 상황에서, 옥상이 사용 주체들에 의해 스스로 관계망을 엮어 나가는 의미 있는 장소가 되도록 고민해야 된다는 것이다. 이것이 우리가 궁극적으로 밟아 나가고자 한 디자인의 발전방향이다. 건축가와 디자이너는 빈틈 없이 완벽하게 조직된 옥상 계획의 욕망을 내려 놓아야 한다. 그 대신 계획 담당자의 자리를 옥상의 주인들에게 일부 내어줘 사용자들 스스로가 필요에 따라 채워나가는 공간이 되도록 해야 한다.

옥상은 생활의 변화를 담는 그릇이 된다. 그릇에는 매일 새로운 먹거리가 채워지고 다시 사라진다. 어제 우리의 옥상은 놀이터였지만, 오늘은 식당이고, 다시 내일은 운동장이 된다. 옥상은 가변적 프로그램으로 채워진다. 가변 공간은 과거에서부터 현재까지 많은 건축가들에 의해 시도되었던 주제였다. 게리 리트벨트는 슈뢰더 주택에서 이동되는 칸막이들을 계획하여 방의 크기가 확장 또는 축소되게 하였다. 알바 알토 역시 부오크세니스카 교회에 가변적인 벽체를 설치해 공간의 크기를 조절했다.

필요하다면 일정부분을 움직여 요구에 대응할 수 있도록 계획한 것이다. 그러나 이러한 가변성은 정해진 몇 가지 시나리오에 대응하기 위한 장치들에 지나지 않았다. 과거의 가변성이 밑그림 위에 반투명종이를 덮고 흔적을 따라 재현을 반복하는 작업이었다면, 오늘날에는 텅 빈 백지(tabula rasa)위에 매일 새로운 그림을 그려나가는 작업이란 것이다. 그렇기 때문에 우리가 상상하는 지금 이 순간의 옥상은 지속적인 변화 중에 순간적으로 모습을 드러냈다 사라져 버릴 단편적 이미지인 것이다. 백지 위에 그림을 그려 나가는 실천은 끊임없는 변용의 잠재력과 대상의 변화를 수용하는 과정이다.

이러한 시도들에도 한계는 존재한다. 자칫 무분별한 기능들이 난립하거나, 예상하지 못한 방향으로 변질될 위험성이 있다. 최악의 경우는 무의미한 공허가 되어 버리는 것이다. 계획을 담당한 건축가나 디자이너에게는, 사용자들의 자발적인 참여를 유도하여 프로그램이 활성화될 수 있는, 기본 토대를 마련하는 역할이 주어진다. 더불어 필요한 장소를 스스로 만들어가는 구성원들에게는 더 강한 결속력 위에서 형성된 관계망이 필요하게 된다. 따라서 어느 누구에게도 필연적 의무는 주어지지 않는다. 다만 더 바람직한 선택의 기회들만이 던져진다.

The rooftop where many owners share together should contain a variety of desires. To do that, the priority should be given to the work which does not add something to, but vacates a given space for the functionality. However, to propose an extreme emptiness as the result of a design can produce only the meaningless no-load rotation. Therefore, an architect or a designer in charge of a planning should be worried so that a rooftop can become a meaningful place where users can weave their networks by themselves in a given situation. This is the development direction of a design which we attempt to take steps in the long run. An architect and a designer should put down insatiable desires about a plan for a rooftop organized perfectly and thoroughly. Instead, they try to create a space where users can fill by themselves according to their needs by giving a position of a person in charge of planning to owners of a rooftop.

A rooftop becomes a bowl which contains a change in life. New food is filled with in a bowl and disappears every day. Our rooftop of yesterday was a play ground, but is a restaurant today, and will again be a schoolyard tomorrow. A rooftop is filled with changeable programs. The changeable space is a topic attempted by many architects from the past to the present. Gary Rietveld expanded or reduced the size of a room by planning movable dividers at Schroder's house. Alvar Aalto adjusted the size of a space by installing a changeable wall in Vuoksenniska Church.

It was planned to respond to the demand by moving a certain part if necessary. This changeability was just one of installations needed to respond to fixed several scenarios.

If the past changeability was the work which the translucent paper was covered on a rough sketch and reproduction was repeated along with traces, the present one is the work which a new picture is drawn on a tabula rasa every day. Therefore, a rooftop of this right moment which we can imagine appeared in a moment among constant changes. It is a fragmentary image which can disappear. The practice which a picture is drawn on tabula rasa is the process which potential of endless change and change of the object are accepted.

There are limitations in these attempts. There is riskiness that indiscreet functions are jumbled up close together or changed into the unexpected direction. In the worst case, it becomes a meaningless emptiness. It is given to an architect or a design in charge of planning, the role in preparing a basic foundation which voluntary participations of users are induced and programs can be revitalized. In addition, the network formed on stronger solidarity is needed to members who create a necessary place by themselves. .
Therefore, inevitable obligation is not given to anybody. Only opportunities for more desirable selections are cast.

2010년 옥상 프로젝트는 '모두를 위한 하나의 공간' 그리고 '다수의 주체에 의해 끊임없이 변화하는 공간' 이란 전제 위에서 새로운 공간을 모색하는 과정이었다. 흥미롭게도 프로젝트의 발전과정 자체와 우리가 구현하고자 고민했던 공간들의 결과가 많은 부분 닮아있었다. 다시 말해 선생과 제자 사이에서 발생하는 영향력에 의해 프로젝트의 발전과정이 구체화 됐듯이, 프로젝트의 결과들은 구성원들 사이의 관계망에 의해 생성과 소멸이 점증적으로 진행되는 공간이 된 것이다.
최종적으로 얻어진 스튜디오의 결과들은 단지 두 개의 얼굴만을 갖고 있는 야누스의 공간에 머물지 않고, 예상치 못한 매개를 통해 끊임 없이 새로운 사건으로 이어지는 앨리스(이상한 나라의 앨리스)적인 공간이기를 희망한다.

나의 의지가 아닌 다른 사람이 매개가 되어 변화하고 있는 나의 모습. 그래서 늘 새로운 나를 발견하게 되는 나. 비록 그것이 내가 그토록 욕설을 퍼부었던 그 사람의 모습일지언정 말이다.
그것이 2010년 옥상 프로젝트에서 실험하고자 한 공간이다.

The 2010 rooftop project was the process which a new space was sought on the premise of 'A Single Space for Everyone' and 'A Endlessly Changing Space by Many Subjects'. Curiously, there were many parts resembled between the development process of the project and results of spaces which we were worried to realize. In other words, results of the project became a space where creation and extinction was gradually progressed by the network among members as the development process of the project was taken concrete shape by influence between a teacher and students.
I hope that results of a studio finally achieved will not just stay at a space of Janus with two faces and become an Alice type (Alice in Wonderland) space where new occasions can constantly keep appearing through unexpected medium.

I want to have my persona which has been changing by other people, not by my will and thus which a new self is always discovered even if it is an appearance of a person who I shout abuses all that.
This is a space attempted to be experimented at 2010 rooftop project.

1st day_07, september ORIENTATION | event hall AM 11:00

1ST day_02 september - ORIENTATION | event hall | AM 11:00

62th day_08, november MID-TERM CRITIQUE | Event Hall PM 14:00 24 WORKS

100th day_ 17. december EXHIBITION

EVENT HALL AM 11:00 23 WOR

ROOF TOP

7.-24. / 22 WORKS, FINAL CRITIQUE & EXHIBITION

OFESSOR

MEBERS

Together Becoming - Kwon Hae-Mi | Kim Hye-Ri | Jang Soon-Hee.

158

100th day 15, december EXHIBITION

Studio_A
김석영　Kim, Seok Yeong

Studio_B
박영태　Park, Young Tae

ROOF
TOP

SCALE1/100

A Aleatorik

'알레아토릭' 이라고 읽으며 '주사위 던지기' 의 의미를 가진다. 이는 예술창작과정의 일부, 혹은 전체를 '우연' 의 놀이에 맡기는 방법으로 우연에서 오는 아름다움을 의미하는 것이다. 여러 가지 프로그램이 만들어지는 해맞이길에서 여러 사람들이 만들어 가는 새로운 길. 이 역시 우연같은 운명이라고 볼 수 있다. 또한 주사위를 던져서 운명을 결정 한다' 는 뜻으로 사람들이 운명적으로 알레아토릭을 선택하게 된다는 의미이다.

Aleatorik means 'Rolling the dice'. This signifies the putting a part or the whole of the artistic creative process in the hands of a 'chance' and represents the beauty which accompanies the concept of chance. The process or putting all in the play of 'chance'. Sunrise Road is where various programs are made and the road which different people newly create. It can be seen as the 'the fate resembling chance'. It also signifies 'rolling the dice and determine the destiny,' meaning that people get to choose Aleatorik by destiny.

 **Aleatorik**

1. 비계획적인 프로그램 : 관객의 참여 예술 (Unplanned Program : Participation art of Audience)
2. 우연에 의거한 제작 태도 (Making attitude is based on coincidence)

· 다다이즘 미술가 마르셀 뒤샹(1887~1968)은 1m 길이의 실 세 개를 떨어뜨려 그대로 작품을 삼았다.
· 초현실주의 조각가이자 시인인 장 아르프(1887~1966)는 글자가 적힌 종이를 찢어 바닥에 떨어뜨려 얻어지는 낱말의 우연한 조합으로 시를 지었다.
· 전위예술가 존 케이지는 샘플을 무작위로 배열해 작곡을 했다.
· Dada artist Marcel Duchamp(1887~1968) :
Three rolls of thread of 1m length were dropped and as such, created a art work.
· Surrealist sculptor and poet hans arp(1887~1966) :
The papers with the words written them were torn into pieces and dropped on the floor and the incidental combination of the words combined to create a poem.
· Avant-garde artists John Cage : Samples were randomly arranged to make a composition.

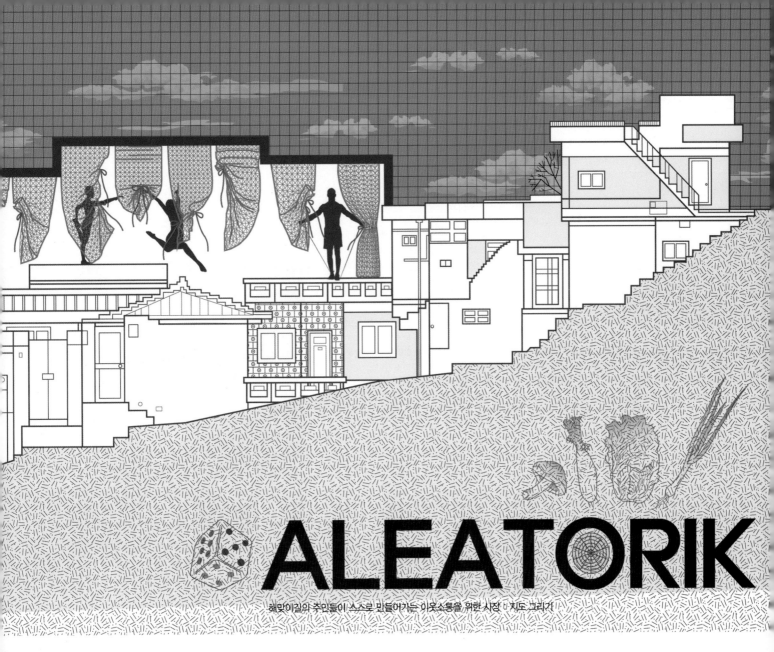

ALEATORIK

해맞이길의 주민들이 스스로 만들어가는 이웃소통을 위한 시장 및 지도 그리기

BACKGROUND

해맞이길은 70년대 이후 도시로 몰려든 탈 농촌 빈민들이 살던 곳으로 40년이 지난 지금도 여전히 불우한 환경이 남아 있는 장소이다. 하지만 이곳 주민들은 고단한 일상의 터인 밀집된 주거의 옥상, 난간 그리고 계단 등의 협소한 공간을 이용하여 채소를 키우고 있다. 그래서 이들이 함께 채소를 키울 수 있는 공간을 옥상에 마련해 주고자 한다. 또한 경작한 채소를 이웃들과 교환할 수 있도록 채소시장을 제안하여 이웃들 간의 의사소통이 활성화 될 수 있도록 한다. 이러한 계획은 해맞이길 주민들에게 자급적인 공동체를 형성하는 동력이 될 수 있을 것이다. 결과적으로 삭막한 분위기의 현대의 도시 한 가운데에 정이 넘치는 공동체 공간으로 발전해 나갈 수 있기를 기대한다.

Sunrise Road is the place of residence for the underprivileged that fled to the city from rural area after the 1970's and after 40 years has passed the place still remains in the disadvantaged environment. However, its residents are using the confined spaces like the rooftops, rails or steps of the densely situated residences where they toil away day after day. So I would like to prepare a space on the rooftop where I can grow vegetables along with these people.

Furthermore, I would like to propose the vegetable market where cultivated vegetables could be exchanged between neighbors, revitalizing the communication between the neighbors. Such plan could prove to be the motivating force behind the forming the self-sufficient community for the residents of the Sunrise Road. I expect it to develop as the community place full of warm affection in the middle of the dreary modern city as a result.

해맞이길은 용산구 한남동에 위치한 산동네이며 동남쪽을 향하고 있으며, 동쪽에서 올라오는 아침 해를 맞이한다는 뜻에서 '해맞이길' 이라고 이름 지어졌다. 이곳은 남산과 한강이 둘러싸고 있는 배산임수의 지형으로, 풍수 지리적으로 볼 때 최고의 명당이라 할 수 있다. 하지만, 해맞이길은 불량 주택의 집합소로, 주택의 형태는1960년대 단층서민 주택에서 1970년대 경제 개발 계획 이후 2~3층의 다세대 주택으로 현재의 모습을 이루고 있다. 해맞이길은 언덕 위에 비계획적인 도시로 형성되었으며, 그 결과 동심원 길과 방사선 길이 서로 만나 격자구도가 약하게 나타나는 불규칙한 동네로, 골목길은 심하게 곡절 되고, 삼거리를 주로 이루며, 좁은 계단과 경사진 길, 탁 트인 시야를 느낄 수 있다.

Haemajigil, Hannam-dong, Yongsan-gu, Seoul, Korea Haemajigil
Sunrise Road, a village located in the hill in Hannam-dong, Yongsan-gu is facing the southeast direction and it was named 'Sunrise Road' meaning that the village greets the morning sun rising from the east. This is the topography oriented in the south direction, surrounded by Nam-san and Han River, the most propitious site from the perspective of feng shui, wind and water. Nevertheless, it is the collection of poorly built houses and the form of residence changed from the single-level affordable housings of the 1960's to the current two to three-stories houses after the Economic Development Plan in the 1970's. From the beginning Sunrise Road was formed as the unplanned city on the hill and as a result, concentric roads and Rhizome roads met and created irregular village weak in lattice structure. The alleys are winded severely and the three-way interactionsare mainly found. With narrow steps and slanted roads, you can see things with wide open view.

LOCAL SURVEY

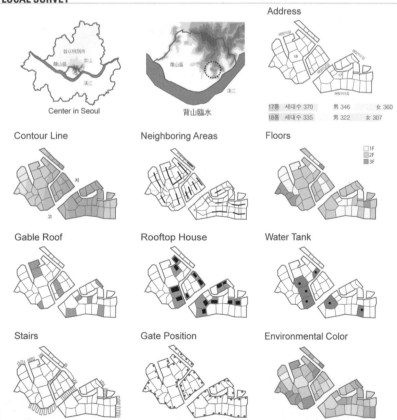

Center in Seoul

背山臨水

Address

| 17통 세대수 370 | 男 346 | 女 360 |
| 18통 세대수 335 | 男 322 | 女 307 |

Contour Line

Neighboring Areas

Floors

☐1F
☐2F
■3F

Gable Roof

Rooftop House

Water Tank

Stairs

Gate Position

Environmental Color

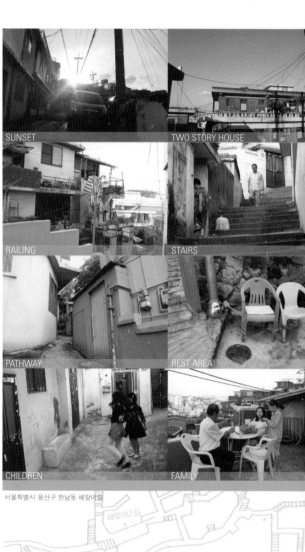

SUNSET

TWO STORY HOUSE

RAILING

STAIRS

PATHWAY

REST AREA

CHILDREN

FAMILY

서울특별시 용산구 한남동 해맞이길

Haemajigil, H

ROOF GARDEN

CONNECTION BIRGE

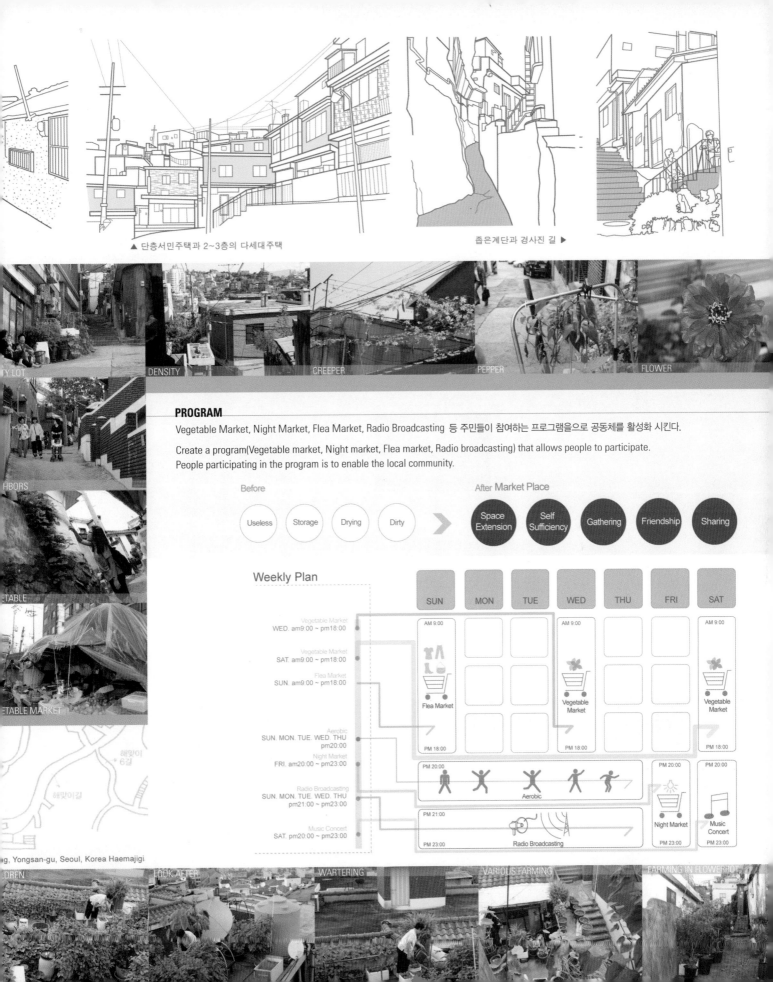

▲ 단층서민주택과 2~3층의 다세대주택

좁은계단과 경사진 길 ▶

VEGETABLE LOT

DENSITY

CREEPER

PEPPER

FLOWER

NEIGHBORS

VEGETABLE

VEGETABLE MARKET

PROGRAM

Vegetable Market, Night Market, Flea Market, Radio Broadcasting 등 주민이 참여하는 프로그램으로 공동체를 활성화 시킨다.

Create a program(Vegetable market, Night market, Flea market, Radio broadcasting) that allows people to participate.
People participating in the program is to enable the local community.

Before

Useless · Storage · Drying · Dirty

After Market Place

Space Extension · Self Sufficiency · Gathering · Friendship · Sharing

Weekly Plan

SUN · MON · TUE · WED · THU · FRI · SAT

Vegetable Market
WED. am9:00 ~ pm18:00

Vegetable Market
SAT. am9:00 ~ pm18:00

Flea Market
SUN. am9:00 ~ pm18:00

Aerobic
SUN. MON. TUE. WED. THU
pm20:00

Night Market
FRI. am20:00 ~ pm23:00

Radio Broadcasting
SUN. MON. TUE. WED. THU
pm21:00 ~ pm23:00

Music Concert
SAT. pm20:00 ~ pm23:00

AM 9:00 — Flea Market — PM 18:00

AM 9:00 — Vegetable Market — PM 18:00

AM 9:00 — Vegetable Market — PM 18:00

PM 20:00 — Aerobic

PM 20:00 — Night Market

PM 20:00 — Music Concert

PM 21:00 — Radio Broadcasting — PM 23:00

PM 23:00 · PM 23:00

Yongsan-gu, Seoul, Korea Haemajigi

CHILDREN

LOOK AFTER

WARTERING

VARIOUS FARMING

FARMING IN FLOWERPOT

CONCEPT

TRACING

해맞이길의 지역적인 특성을 조사한 결과들을 가지고 우연한 공간을 만들고, 공간의 형태는 그 지역 주민들과 오고 가는 행인에 의해서 불확정적으로 만들어지는 공간으로서, 시간에 따라, 오는 사람에 따라, 파는 물건에 따라 달라질 수 있다.

With the results from examining the regional characteristics of the Sunrise Road and created a coincidental space. The form of the space is created with uncertainty by the coming and going of the pedestrians and the residents of the area, which changes according to time, people coming and the goods sold.

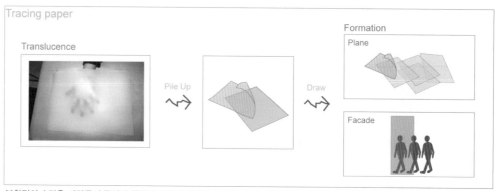

불확정의 수많은 사람들의 동선과 행위에 의해서 우연한 경우의 파사드와 공간이 결정된다.

Due to the flows and actions of uncertain number of countless people, facade and the space have become determined accidentally.

Touching Transparent Fabric

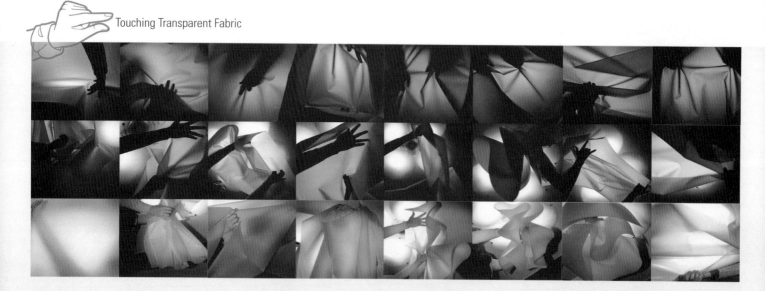

MARKET PLACE PROCESS

1. 근접도를 바탕으로 한 해당영역 위에 박공지붕은 제외
2. 박공지붕을 제외하고 남은 영역에 옥탑방 제외
3. 다른 영역과 비슷한 레벨사이에 경로 설정
4. 경로로 이을 수 없는 영역을 삭제하고 남은 영역이 최종 장터 영역

Based on the proximity remove the gable roof from the given area and take away the attic room from the remaining area after removing gable roof.

Then set the route between the similar levels with other areas and remove the areas where routes cannot be connected and the remaining area is the final market site.

PROCESS

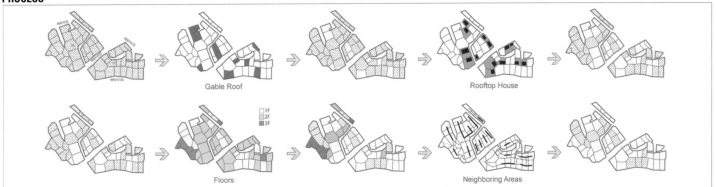

Gable Roof   Rooftop House

1F
2F
3F

Floors   Neighboring Areas

Market place   Farming place

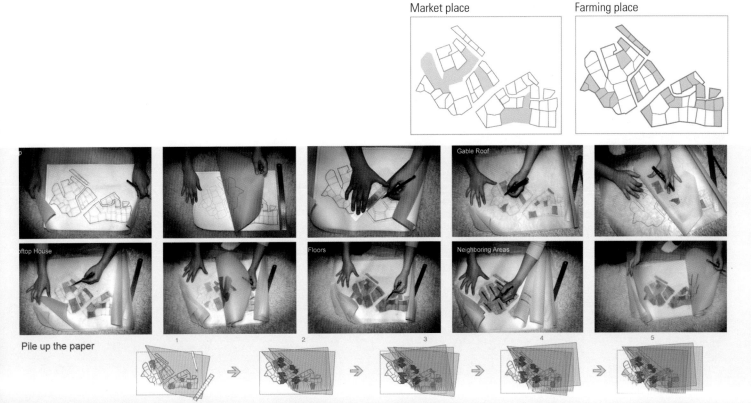

Gable Roof

Rooftop House   Floors   Neighboring Areas

Pile up the paper

1   2   3   4   5

FORM PROCESS

PHASE

no.1

no.2

no.3

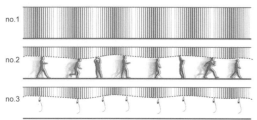

ELEVATION

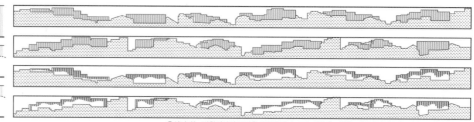

옥상이라는 비활성화된 공간을 사람들이 스스로 커튼의 모양을 만들면서 필요한 공간으로 활성화된다.
The rooftop, which is the space disused, is used as the necessary space with the various shapes of curtains made by the people on their own.

FABRIC GRID
가로, 세로, 대각선의 그리드를 바탕으로 실을 끼워 커튼을 완성한다.

1000×2600

사람들은 커튼을 당겨서 원하는 모양으로 만들 수 있다.
The people can pull the curtain and make it into the shape they want.

IRREGULAR FORM

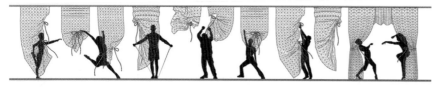

ACCDIENTAL FORM

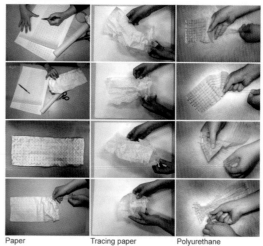

Paper   Tracing paper   Polyurethane

GRID FRAME
커튼을 매달 수 있는 트러스 구조를 격자모양으로 설치한다.
Based on the horizontal, vertical and diagonal grid, the curtain was completed by threading.

Curtain   String   Hand   Touch   Shape

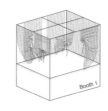 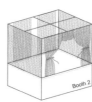  

Booth 1   Booth 2   Booth 3   Booth 4

MANNER PROPOSE

Marketplace

Traffic Line

Making Space

Tuck up

According to the need

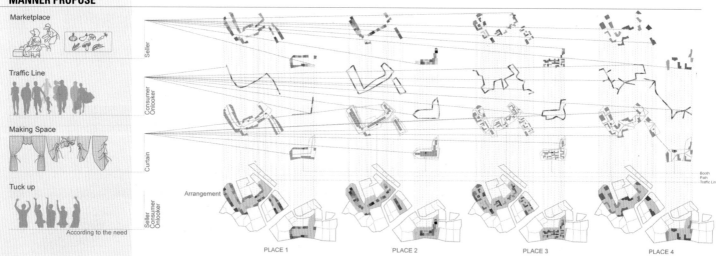

Seller

Consumer
Onlooker

Curtain

Seller
Consumer
Onlooker

Arrangement

PLACE 1   PLACE 2   PLACE 3   PLACE 4

PROGRAM SUBDIVISION

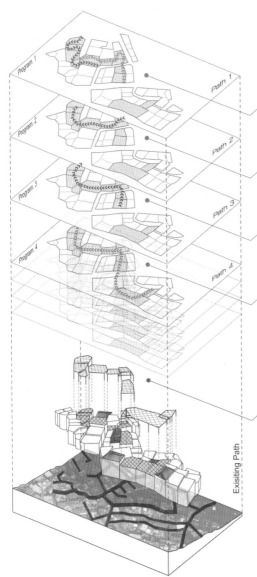

Program. 1 — Path 1
Program. 2 — Path 2
Program. 3 — Path 3
Program. 4 — Path 4

Exisiting Path

Vegetable Market

Night Market, Flea Market

Public Stadium, Rest, play

etc : Radio Broadcasting, Family Party

Grid Frame

Rooftop

한남동 해맞이길

Farming Place
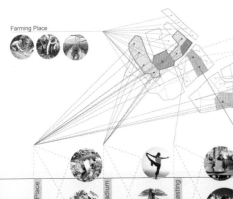

Aerobics
Music Concert
Farming Place

Market Place	Public Stadium	Radio Broadcasting	Market Square	Family Party
Vegetable Market	Rest	Impartment	Center	Camping
Flea Market	Play	Play Song	Have a Talk	Barbecue
Night Market	Aerobics	Communication	Drink	Cooking
	Jogging	News	Social get-together	Amusement
	Gymnastics	Meeting	Break Time	
	Health		Music Concert	
	Walk			

VEGETABLE MARKET

해맞이길

Radio Broadcasting

Aerobic, Gym

Play, Meeting

Jogging

Search One's Memory

_ Memory of the life carved on the ground

급격한 도시화와 인구증가, 인구 밀집현상으로 인해 한정된 공간에서 지나치게 많은 사람들이 살고 있다. 한정된 공간속에 밀집될수록 사람은 폐쇄적으로 변하며 심각한 사회문제의 기반이 되고 있다.
우리는 다세대 주택단지의 폐쇄성 해결을 위해 일상 속에서 이웃과 함께했던 옛 우리 삶의 기억과 추억을 되찾고자 한다

Based on hasty urbanization, population increase, and overpopulation phenomenon, overly many people live in limited area. More concentrated in the limited area, people become unsociable and it becomes a base of serious social problems. To solve the closing, we want to retake our memory of life and reminiscences with neighbors in routine days.

STORYBOARD

19세기초 도시화로 인해 지방에 있는
사람들이 모두 도시로 올려왔었다.
In early 19c, for urbanization, people in
local area all came to cities.

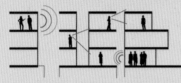

사람들이 모여들어 공간이 부족하자 좁은 곳에
밀집해서 살기 시작했어. 그러자 많은 문제가
발생한거야.
The people started to live in narrow area when they
didn't have enough space. Then, many problems
happened!

옆집에서 들리는 시끄러운 소음과 가까운 주택간의
거리로인한 사생활침해로 사람들은 이웃과도 멀어지고
좁은공간에 자기 자신을 가두기 시작했어.
밀집되고 폐쇄적인 주택들로인해 스트레스와 반사회적 행동도
늘어나게 됐지. '인간동물원'현상이 나타나는거야.

* 인간동물원화
동물들이 갇혀 있기 전에는, 스트레스나 무차별적인 폭력, 변칙적인 행동을
드러내지 않는다. 마찬가지로, 인간의 반사회적 행동양식은 오늘날의 밀집되고
폐쇄적인 도시 안에서 더욱 격렬해졌다.

Because of noisy sounds from the next door and narrow distance between houses, people drifted apart and started to
take themselves into the small space.
With concentrated and closed houses, stress and antisocial
actions also got increased. 'Human zoo' phenomenon appeared!

*human zoo phenomenon
Before animals get trapped, they don't show stress, indifferent vio-
lence, and irregular behaviors. Likewise, human's antisocial behavior
patterns became more violent inside of the concentrated and closed
cities.

이러한 일들이 반복되면서 사람들은 서로 마음의
문을 닫게 되었지. 폐쇄적이고 밀집된 도시에서 살지
않는 사람들은 어떻게 지낼까?

As these events happened again and again,
people closed their heart each other. How did
people in the concentrated and closed cities
live?

아버지 말씀이 예전에는 서로를 많이 도와주며 이웃들과
가족처럼 가깝게 지내고, 자연과 친해지는 법을 어른들에게
배웠대.
My father said that people helped others a lot and
maintain their relationships friendly, and learned
about how to be familiar with nature from adults.

우리 동네도 이렇게 변화될 수 있지 않을까? 그런 활동을 할 수 있는
공간을 지금은 잘 활용하지 않는 옥상을 이용해보는 것은 어떨까?
Our town also couldn't be changed like
this? How about use a rooftop not used
often where can do those activities?

SITE ANALYSIS

서울 구로구 개봉동 90-30 (청풍길 30-10)
90-30 Gaebong-dong Guro-gu, Seoul (Chungpung-gil 30-10)

etc
Because nasty stuffs are loaded
Because the view is not good
Because there is no resting place
Because nothing special to go up
Because not interested in the rooftop
Because people around there matters
Because someone lives on the rooftop
Because the rooftop door locked
Because the only top floor residents can use

**The reasons why multifamily house residents' don't use the space of
the rooftop (multiple choices possible)**

최상층세대만 사용하게 되어있어서
옥상공간 출입문이 잠겨있어서
옥탑에 사람이 거주하고 있어서
주변사람들이 신경쓰여서
옥상에 관심이 없어서
특별히 올라갈일이 없어서
쉴만한 공간이 존재하지 않아서
주변의 풍경이 좋지 않아서
지저분한 물건들이 적재되어 있어서
기타

다가구주택거주자들이 옥상공간을 이용하지 않는이유 (중복선택가능)

site
the southern circuit
underpass

SITE 왼쪽으로 남부순환로가 위치하며 도로 아래 굴다리를 통
해 사람들의 이동이 주로 이루어지고 있다.
SITE On the left the southern circuit is located, people
pass through the underpass under the road.

버스정류장이 남부순환로 위에 위치하고 있다는 점을 이용하
여 버스정류장에서 바로 옥상으로 진입로를 연결한다.
From the point that a bus-stop is placed on the southern
circuit, connect the entrance from the bus-stop to the
rooftop directly.

CONCEPT

급격한 도시화와 인구증가, 인구 밀집현상으로 인한 밀집된 주택에서 발생하는 다양한 문제점을 해결하기 위해 일상 속에서 이웃과 함께했던 옛 우리 마을의 추억을 옥상위로 올려 보고자 한다.

To solve the various problems caused by hasty urbanization, population increase, and overpopulation phenomenon, I want to lift our town's reminiscences up to the rooftop.

터무늬(地文)

우리 모두에게 각자 다른 지문이 있듯이 모든 땅도 고유한 무늬를 가지고 있다. 더러는 자연의 세월이 만든 무늬이며, 더러는 그 위에 우리의 삶이 연속적으로 새긴 무늬이다. 도시화로 인해 오랜 삶의 터들은 사라져야 했으며 산이 있으면 깎고 계곡의 물길은 왜곡되어야 했다.

터무늬(地文)는 우리가 이제껏 해왔던 모든 역사, 추억을 지니고 있는 과정, 결과물로써 현대사회에서 오는 폐쇄적인 문제점을 해결해줄 수 있는 방안을 과거에서 찾을 수 있다.

LANDSCRIPT(Site pattern)_As we have our own unique fingerprint, every site has its own special patterns. Some are made by the time of the nature, the other are carved on them by our lives continually. For the urbanization, sites of long-standing lives had to be disappeared, the mountain had to be cut and the streams had to be distorted.

The site patterns can find all the history we have done, process having reminiscences, and a plan that can solve the closed problems from the modern society as a result in the past.

품앗이

폐쇄적인 문제를 실제로 이루어지는 행동이나 직접적으로 느낄 수 있는 행위로 해결. 작은 힘들을 모아 이웃에 힘든 일이 있을 때 서로 거들어 주면서 품을 지고 갚고 했다.

"PUMASI"(=Exchange Of Labor - Korean Traditional) _ Solve the closed problems with actual behavior or action that could feel directly. When neighbors have something hard to do, people gathered and helped each other.

분합문

폐쇄적인 문제를 분합문의 물리적 요소로 해결. 분합문을 이용하여 날씨의 변화나 공간의 필요에 따라 가변적인 공간을 형성하였다.

"BUNHAPMUN"(Korean Traditional Sliding Doors) _ Solve the closed problems with sliding doors' physical factor. Using them, form changeable space as the change of the weather or needs of space.

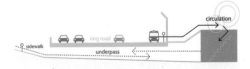

줄어든 동선으로, 옥상은 항상 많은 계단을 올라가야만 갈 수 있는 곳이라는 고정관념을 깨고 옥상과 보도의 경계를 모호하게 하여 자연스럽게 옥상으로 사람들을 끌어들일 수 있다.
With reduced movement, after breaking stereotype that the rooftop is the place where we always think we should go up many stairs, we can make the border between the rooftop and sidewalk ambiguously and attract people naturally to the rooftop.

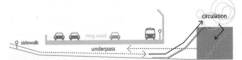

두 지역을 잇는 유일한 통로인 굴다리를 옥상으로 연결된 진입로는 더 많은 사람들을 끌어들인다.
The entrance connected with the rooftop that is connected with the underpass which is the only pathway connects two areas attracts more people.

SPACE FUNCTION

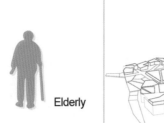

Elderly

Kids

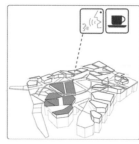

Everyone

CONCEPT PROCESS
process -1

터무늬(地文)를 읽어내어 지켜야 될 것과 변화해야 할 부분을 판단하고 땅의 기억을 바탕으로 그 안에 필요한 기능들을 성질에 맞게 얹는다.

Read a LANDSCRIPT(site pattern) and judge which part is should be preserved or changed, and assign some necessary functions on it based on memory of the ground.

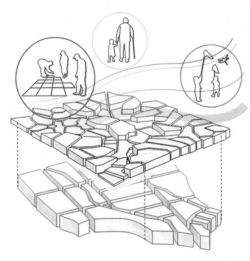

개봉동(Gaebong–dong)

급진적인 도시화, 현대화로 인한 획일화된 땅의 모양

Rapid urbanization, unified ground shape because of the modernization.

옛 마을(Old town)

마을에서 끌어올린 터무늬는 그 동안의 시간의 축적에서 나타난 우리 삶의 추억과 기억들을 지님.

A site pattern dragged from the town contains our reminiscences of life shown from previous time accumulation.

이성적이고 획일화된 건물의 옥상위에 옛 우리마을의 터무늬와 시간의 축적에서 나타난 추억과 기억을 올린다.

Lift up reminiscences and memory from our old town's site pattern and time accumulation to rational and unified building's rooftop.

process -2

지켜야 할 것 = 옛 마을의 기억, 추억(실질적 공간)
Things to preserve
= old town's memory, reminiscences (practical space)

기억A
어린이 자연체험공간
Memory A
Kids nature experience space

기억B
어린이+노인
Memory B
Kids + Elderly citizens

기억C
마을입구(노인휴식)
Memory C
Town entrance (elderly citizens' rest)

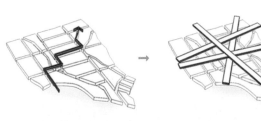

획일화된 도시의 딱딱한 거리를 자유로운 동선으로 변화
Change the unified hard road of the city to free movement.

변화해야 할 것 = 획일화된 동선(이동 공간)
Things to be changed
= unified movement (movement space)

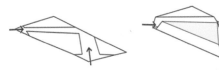

터무늬의 모양 그대로를 따라가며 기억을 찾을 수 있는 방식으로 실질적공간은 레벨을 높여 공간 활성화를 돕고 그를 감싸는 이동공간은 상대적으로 레벨을 낮춘다.

Then following the site pattern and the way that can find the memory, the practical space raises its level and help the space activation, and movement space which wraps it lowers its level relatively.

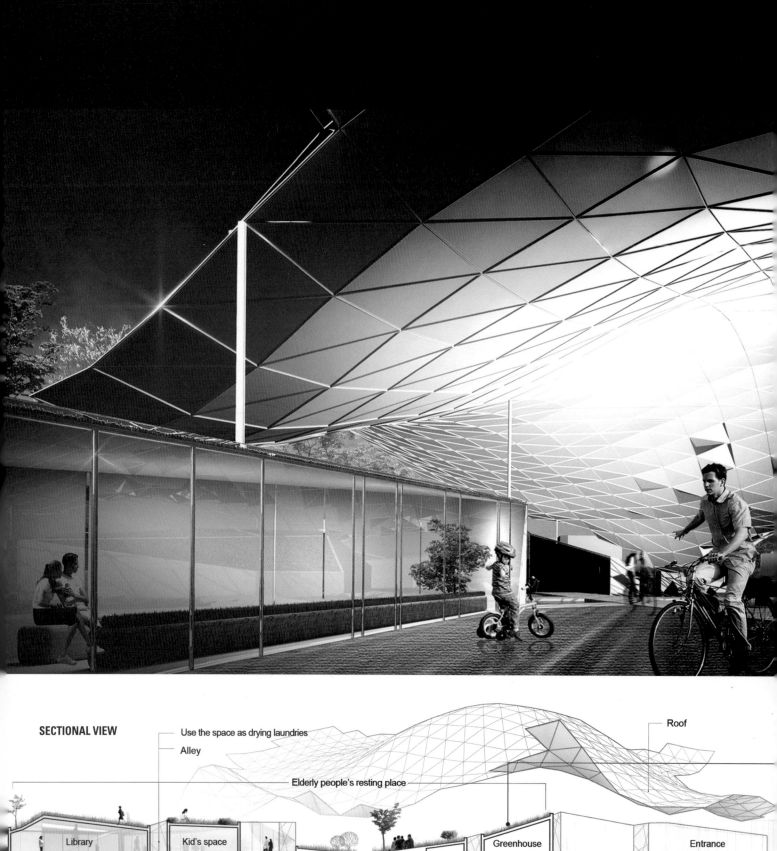

SECTIONAL VIEW

Use the space as drying laundries

Alley

Roof

Elderly people's resting place

Library

Kid's space

Greenhouse

Entrance

Multiple c

ROOF DETAIL FUNCTION

Rotation principle

Dew collection

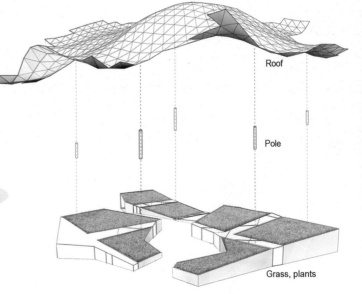

Roof

Pole

Grass, plants

Roof module 작은 물방울들이 모여 큰 효과

비가오거나 새벽에 이슬이 맺혔을 때 물방울들이 roof의 낮은 곳으로 관을 통해 이동하고 각각의 기둥으로 흘러 온실공간과 잔디에 물을 공급한다.

Roof module Great effect from little water drops' gathering
When raining or the dew is formed early in the morning, water drops go through a pipe on the roof's low place and flow on each column and provide water to greenhouse area and grass.

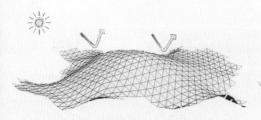
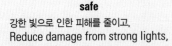

safe
강한 빛으로 인한 피해를 줄이고,
Reduce damage from strong lights,

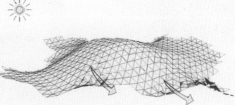

fresh
Roof의 부분을 열어 적절한 채광과 통풍을 가능케 한다.
and open the part of the roof to get proper amount of light and get ventilated.

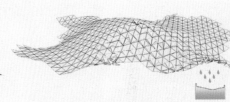

assemble
Roof를 모두 다 열었을 경우에는 비나 새벽의 이슬을 모이게 할 수 있다.
In case of opening the whole roof, rain or early morning dew can be collectable.

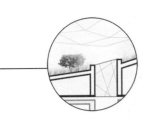

빨래를 너는 공간으로 사용
Use the space as drying laundries.

기존의 주택 사이공간에 유리를 사용하여 채광과 통풍 문제 해결
Existing house gap distance light and ventilation problem solution using glass.

많은 사람들이 옥상을 이용하면서 이 유리를 통해 범죄에 취약한 공간에서는 모든 사람들이 감시자 역할을 함.

바닥의 레벨차를 이용하여 아이들에게 활동적인 활동을 할 수 있도록 하며 가변적인 공간을 스스로 만들어 낸다는 것에 의의를 둔다.

아이들은 각자의 개인 화분을 소유하여 온실공간에서 관리를 한다. 서로 돌아가면서 모두의 화분에 물을 주는 방식으로 화분 관리가 된다면 예전 품앗이라는 공동체 의식이 더욱 발전 될 것이다.

옛 마을 입구에서의 느티나무 분으로 노인들의 쉼터 공간 착안. 벽을 차양 막으로 활용

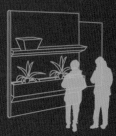

I attach meaning on let children take active actions and make changeable space itself.

Children have their own personal flowerpot and manage the greenhouse area.If they take rotating turn to water their flowerpots, their community spirits previously called Pumasi will be much developed.

This is elderly people's resting place reproducing the view of under the zelkova tree at the old town entrance. Regarding to the slide doors (Bunhapmun) can be lifted, I utilized the wall as an awning.

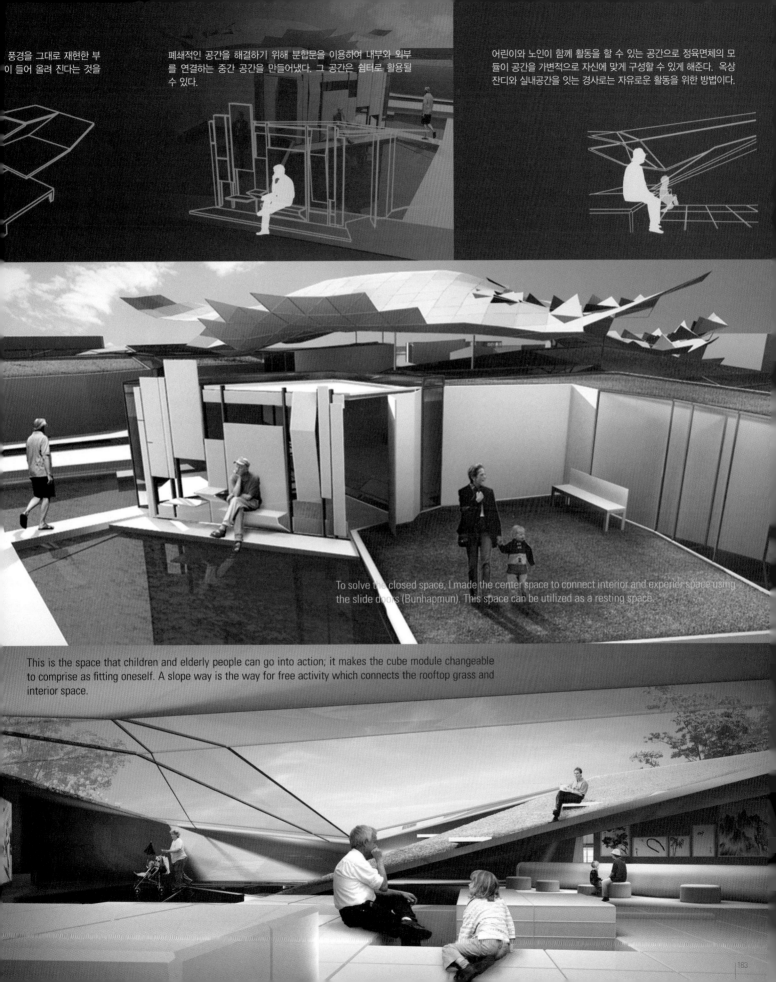

풍경을 그대로 재현한 부
이 들어 올려 진다는 것을

폐쇄적인 공간을 해결하기 위해 분합문을 이용하여 내부와 외부
를 연결하는 중간 공간을 만들어냈다. 그 공간은 쉼터로 활용될
수 있다.

어린이와 노인이 함께 활동을 할 수 있는 공간으로 정육면체의 모
듈이 공간을 가변적으로 자신에 맞게 구성할 수 있게 해준다. 옥상
잔디와 실내공간을 잇는 경사로는 자유로운 활동을 위한 방법이다.

To solve the closed space, I made the center space to connect interior and exterior space using the slide doors (Bunhapmun). This space can be utilized as a resting space.

This is the space that children and elderly people can go into action; it makes the cube module changeable to comprise as fitting oneself. A slope way is the way for free activity which connects the rooftop grass and interior space.

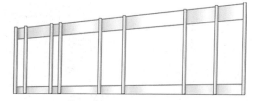

분합문의 활동이 일어날 수 있는 범위 안에서 프레임의 모듈을 정한다.

Frame module should be decided in the field where the sliding doors' activity can happen.

Sliding_ 슬라이딩 방식은 폐쇄적인 공간을 개방적인 공간으로 바꿀 수 있는 역할을 하며 이러한 다양한 방식의 슬라이딩이 공간의 효율성을 돕는다.

Sliding_ A sliding way takes role as it can change closed space into open space and raises efficiency of various ways of sliding.

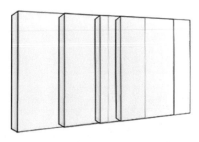

그 모듈안에 적용되는 분합문들은 각각 너비, 높이 등이 달라 다양한 용도의 공간을 형성하는데에 도움을 준다. 이러한 분합문들을 반복하여 배치하게 되면 면적으로도 흥미로운 결과가 나온다.

The sliding doors applied in the module help to form various use of space as each door's length, height and others are different. If place these sliding doors repetitively, an interesting result in elevation comes out.

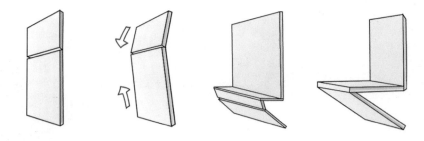

Folding_ 폴딩은 용도가 다른 공간에 따라 가변적으로 적용되어 새로운 공간을 형성할 수 있다. 접히는 방식이나 방향에 따라 다양한 형태를 만들기도 한다.

Folding_ Folding can organize a new space as it is applied changeably regarding different function spaces. It sometimes makes a folding way or various forms as its direction.

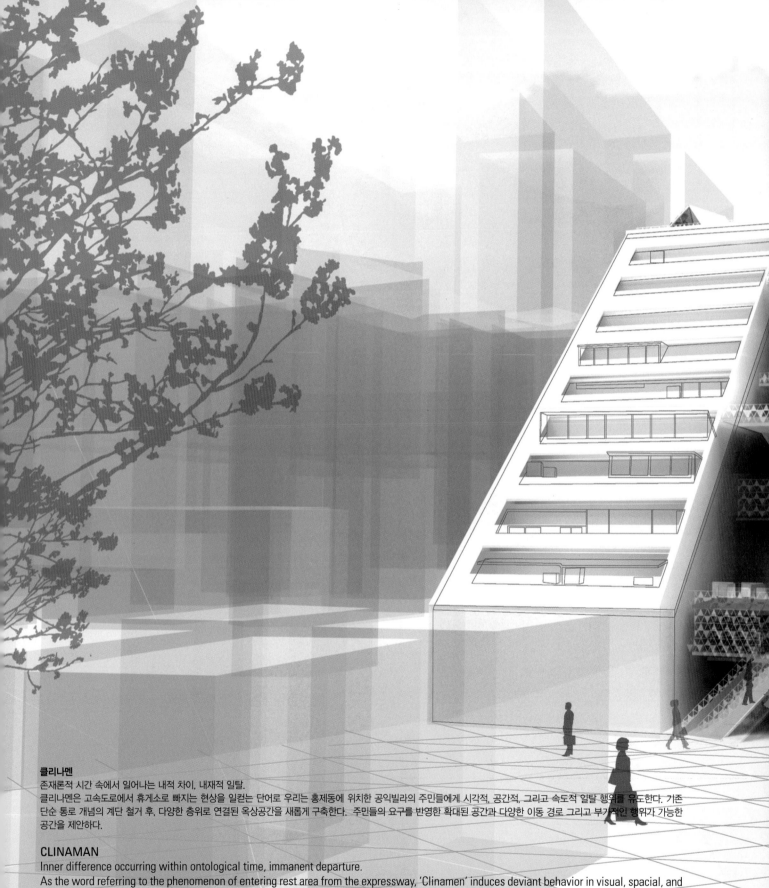

클리나멘

존재론적 시간 속에서 일어나는 내적 차이, 내재적 일탈.

클리나멘은 고속도로에서 휴게소로 빠지는 현상을 일컫는 단어로 우리는 홍제동에 위치한 공익빌라의 주민들에게 시각적, 공간적, 그리고 속도적 일탈 행위를 유도한다. 기존 단순 통로 개념의 계단 철거 후, 다양한 층위로 연결된 옥상공간을 새롭게 구축한다. 주민들의 요구를 반영한 확대된 공간과 다양한 이동 경로 그리고 부가적인 행위가 가능한 공간을 제안하다.

CLINAMAN

Inner difference occurring within ontological time, immanent departure.

As the word referring to the phenomenon of entering rest area from the expressway, 'Clinamen' induces deviant behavior in visual, spacial, and speed aspects for residents of public villa located in Hongje-dong. It newly constructs rooftop space connected by various layers after removing existing stairs of simple passage concept. Expanded space that facilitates diverse moving routes and additional actions is suggested in reflection of the requests of residents.

CLINAMAN

클리나멘 시각, 공간, 속도의 이탈

홍제동 공익빌라, 주민들의 시각적, 공간적, 속도이탈을 유도하는 옥상과 자율적 경로제공

GONGICK VILLA

홍제동 공익빌라

45도 경사지에 지어진 건물은 엘리베이터가 설치되어 있지 않다. 주민들이 출입을 위해 사용할 수 있는 수직 동선으로는 기능적 합리성에 근거한 단순형태의 계단만이 존재한다. 25년 전에 완공 된 이 건물은 상당 부분 노후됐을 뿐만 아니라 주민 커뮤니티를 위한 공간도 요구되고 있다. 더불어 어린이나 노인을 위한 안전한 놀이공간과 여가를 위한 여백의 공간이 절실한 실정이다. 또, 각 주거마다 마련되어 있는 테라스는 그 기능을 충분히 수행하지 못하고 있어 계단실의 상당부분이 빨래 건조대로 사용되고 있다.

Elevators are not installed in buildings constructed on -45 degree slope. Simple-form stairs based on functional rationality exists as the only vertical channel used for entrance by residents. The building that was completed 25 years ago is not only deteriorated to a large degree but also needs space for resident community. Furthermore, safe playing space and marginal space for leisure are in desperate need for children and elderly residents. Terraces arranged in each residential area have not sufficiently executed its function, and are thus being used as clothes dryers.

CONDITION

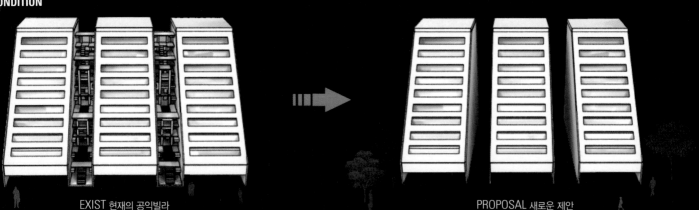

EXIST 현재의 공익빌라   PROPOSAL 새로운 제안

SITE ANALYSIS

Seoul Seodaemoongu

House facing EAST

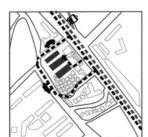

Traffic

Walkways

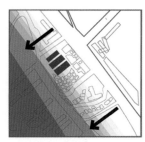

Level

REQUIREMENT

같은지역에서의 생활
Living in the same area

공동생활과 경험의 공유
Sharing of common life and experience

커뮤니티 의식(공동가치의 추구)
The pursuit of common values

커뮤니티 형성
Create Community

NEED

UMBRELLA

WALK & SIT

VEGTABLE GARDEN

PALYGROUND

HANG OUT THE LAUNDRY

BICYCLE RACK

VENTILATION

BACKGROUND

테라스 하우스는 우리집 옥상, 윗집의 테라스가 되는 개념이다. 우리는 이 개념을 공익빌라의 계단에 적용하여 계단이 단지 지붕위에서 뽑아낸 의무적인 시스템이 아니라 계단의 윗 공간이 계단의 옥상이 될 수 있다고 생각한다. 이 옥상은 집합적이면서 개인적인 공간을 제공한다. 수직으로 단순하게 반복되는 계단은 주민들이 갖고 있는 다양성을 수용하지 못하고 일률적인 동선만을 제공하기 때문에, 본 프로젝트의 목적은 기존의 계단실을 철거한 후 확장된 옥상 개념의 계단실을 새롭게 구축하는 것이다.

그 결과로서 주민 편의시설 및 커뮤니티 공간을 제공하는 것이다.

Terrace house refers to the concept of the house rooftop and the upstairs terrace. This concept is applied to the stairs of gonick villa.Stairs just plucked off the rooftop of the stairs, not a mandatory system, space may bethe roof of the stairs, I think. Collective, private space is available. Stairs that are repeated in simple vertical form cannot acommodate the diversity possessed by residents and only provides uniform channel. The purpose of this project is to newly construct expanded staircase in rooftop concept after demolishing the existing staircase. This is constructed to provide resident amenities and community space.

CONCEPT

시에르핀스키 삼각형 **Sierpinski triangles**

경우의 수 Number of cases	2	6	120	. . .	다양한 경로 제공 Provides a variety of paths
삼각형의 개수 The number of triangles	1	3	13	. . .	다양한 공간 제공 Provides a variety of space

1. 정삼각형 하나에서 시작한다.
2. 정삼각형의 세변의 중점을 이으면 원래의 정삼각형 안에 작은 정삼각형이 만들어진다. 이작은 정삼각형을 제거한다.
3. 남은 정삼각형들에 대해서도 2를 시행한다.
4. 3을 무한히 반복한다.
삼각형 하나에서 시작. 삼각형의 세변의 중점에서 이음. 삼각형 안에 작은 삼각형 형성.
반복 되면서 삼각형 안에서의 경우의 수가 형성된다.

1. Start with one regular triangle.
2. Connecting central points of the three sides of regular triangle forms small regular triangle within the original regular triangle. This small regular triangle is removed.
3. 2 is exccutcd on remaining regular triangles.
4. 3 is infinitely repeated. Begin with one triangle. Connect on central point of the three sides of the triangle. Number of cases is formed within triangle with the repetition of small triangle formation within triangles.

삼각형이 많아질수록 경우의 수가 늘어나는 것으로부터 다양한 경로와 그 안에서 형성되는 다층적 공간의 영감(inspiration)을 얻는다. 경로와 공간적 프로그램에 따라 다양한 기능과 경로를 선택할 수 있다. 또자연스러운 동선유도로 다양한 행위를 유발한다. 시에르핀스키 삼각형이 적용된 반복 형태의 계단은 최하층으로부터 수직으로 건물을 관통한다. 그러나 여러 층위의 데크와 경사로는 기존에 수용할 수 없었던 행위들을 수용한다.

KEYWORD

시에르핀스키 삼각형
Sierpinski triangles

축소와 확장 Expand, diminution	자기복제 Replication	면, 선, 각도 Side,Line,Angle	다양성 Multivalence	결합 Combination

Increased number of triangles increase number of cases to gain various routes and inspiration of multi-layered space that are formed inside. Diverse functions and routes can be selected according to spacial program. Furthermore, various actions are aroused through natural channel inducement. Stairs of repeated form applied with Sierpinski triangle vertically penetrates building from the lowest layer. However, various layers of deck and slope accommodate behaviors that had not been accepted previously.

PROCESS 1

다양성
Diversity

라인으로 도출
Lines conclusion

다양성/ 시에르핀스키의 기본 모듈에서 경우의 수를 찾아 그 경로를 형태화시켜 공간을 형성 하고자 하였다.

Diversity/ Number of cases is found in basic module of Sierpienski to form route to create space.

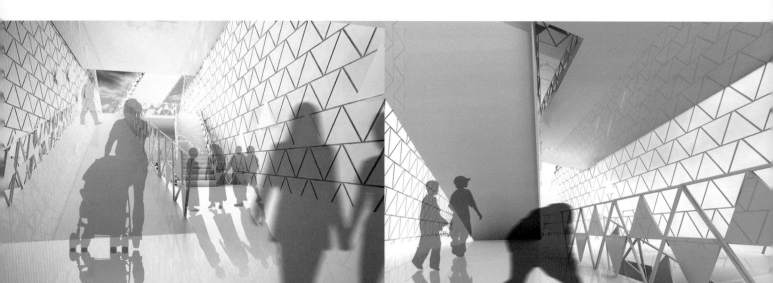

01 경사로와 계단의 길을 주민들이 선택한다.
Slope and stairway are selected by residents.

02 계단 밑 공간을 이용하여 아이들의 놀이공간을 제공한다.
Space under stairs provides playing space for children.

PROCESS 2

삼각형의 축소 Triangle Reduction	반복되는 모듈 복제 Repeat the module replicated	면과 라인의 형성 Sides and lines formation	3개의 라인 도출 3 lines conclusion	도출된 라인의 결합 The combination of line

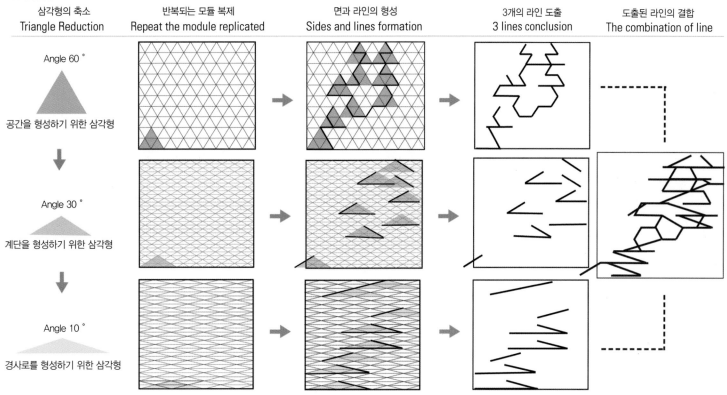

Angle 60°
공간을 형성하기 위한 삼각형

Angle 30°
계단을 형성하기 위한 삼각형

Angle 10°
경사로를 형성하기 위한 삼각형

PROCESS 3

반복 / 시에르핀스키의 기본모듈을 반복 시켜 그리드를 형성
Repetition / Basic module of Sierpienski is repeated to form grid.

Base Module   01 Stairs   02 Lighting   03 Railing   04 Pollen

01 계단공간에 적용하여 의자 형성
02 그리드를 바탕으로 하여 삼각형 조명을 만들어 이동에 도움을 준다.
03 난간에 적용하여 미적효과를 부여 한다.
04 난간과 화분 이중의 역할을 하여 공간의 효율을 높인다.

01 It is applied to staircase space to form chair.
02 Triangle lighting is created based on grid to assist movement.
03 It is applied to banister to grant aesthetic effects.
04 It takes the dual role of banister and flowerpot to increase efficiency of space.

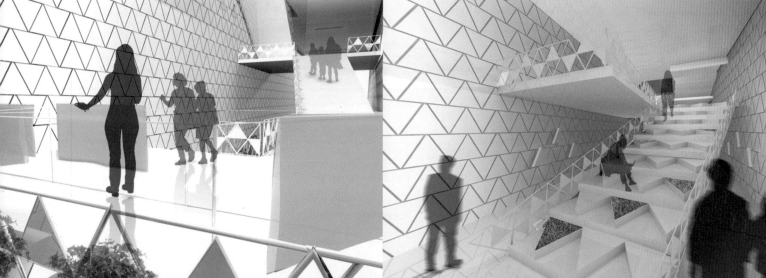

00 계단 공간의 넓은 합을 이용하여 빨래를 널 수있도록 한다.
Wide space of stairs is used to hang laundry.

04 계단이 의자역할을 하여 주민들의 커뮤니티 공간을 형성한다.
Stairs are used as chairs to form community space for residents.

191

SPACE DEVELOPMENT

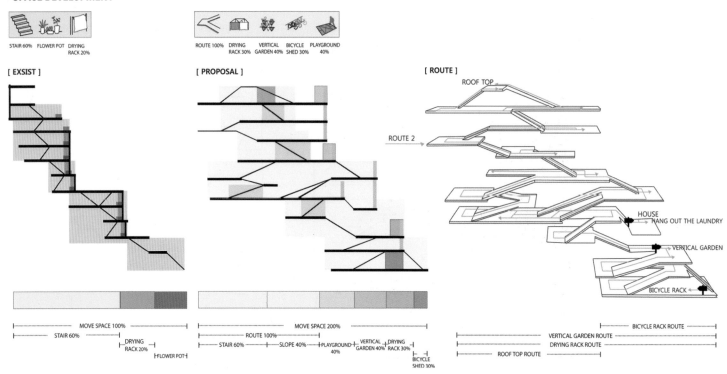

[EXSIST]

MOVE SPACE 100%
STAIR 60%
DRYING RACK 20%
FLOWER POT

STAIR 60%   FLOWER POT   DRYING RACK 20%

[PROPOSAL]

ROUTE 100%   DRYING RACK 30%   VERTICAL GARDEN 40%   BICYCLE SHED 30%   PLAYGROUND 40%

MOVE SPACE 200%
ROUTE 100%
STAIR 60%   SLOPE 40%   PLAYGROUND 40%   VERTICAL GARDEN 40%   DRYING RACK 30%
BICYCLE SHED 30%

[ROUTE]

ROOF TOP
ROUTE 2
HOUSE
HANG OUT THE LAUNDRY
VERTICAL GARDEN
BICYCLE RACK

BICYCLE RACK ROUTE
VERTICAL GARDEN ROUTE
DRYING RACK ROUTE
ROOF TOP ROUTE

PROGRAM

다양한 경로가 만드는 옥상공간으로 공간을 다양하게 구성하여 주민이 이동하면서 사용할 수 있도록 하였다.
Space is diversely composed through rooftop space formed by various routes to enable residents to use while moving.

■ ROUTE   ■ VERICAL GARDEN   ■ PALYGROUND   ■ DRYING RACK   ■ BICYCLE RACK   ■ COMMUNITY

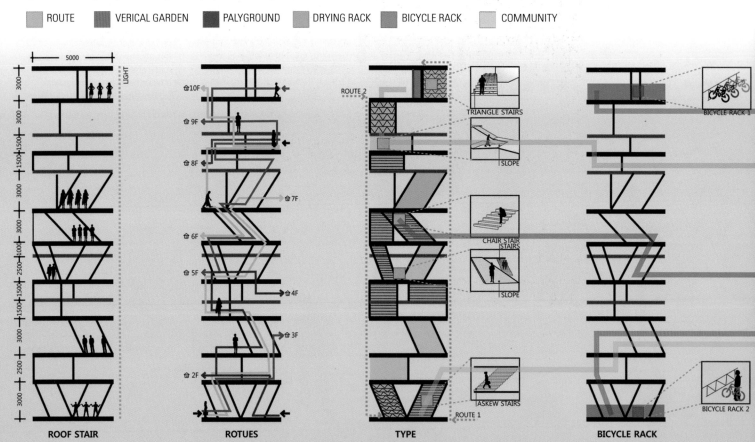

ROOF STAIR   ROTUES   TYPE   BICYCLE RACK

ROUTE

Route 1

Route 2

Route 3
부통로2   부통로1   주요통로
현관문

Route 4
계단   경사로

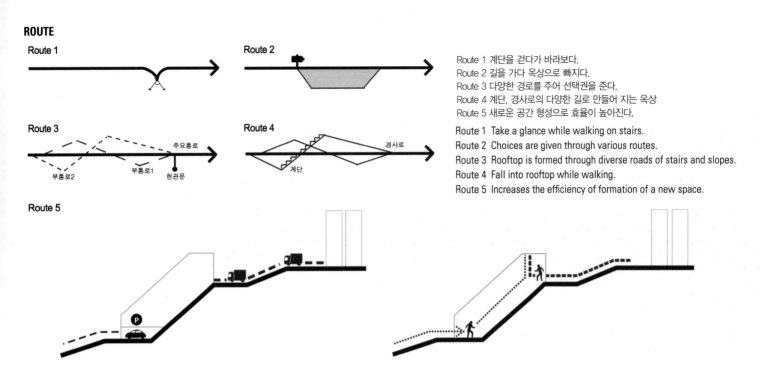

Route 1 계단을 걷다가 바라보다.
Route 2 길을 가다 옥상으로 빠지다.
Route 3 다양한 경로를 주어 선택권을 준다.
Route 4 계단, 경사로의 다양한 길로 만들어 지는 옥상
Route 5 새로운 공간 형성으로 효율이 높아진다.

Route 1   Take a glance while walking on stairs.
Route 2   Choices are given through various routes.
Route 3   Rooftop is formed through diverse roads of stairs and slopes.
Route 4   Fall into rooftop while walking.
Route 5   Increases the efficiency of formation of a new space.

Route 5

ROUTE 계단, 경사로, 앉을 수 있는 계단으로 구성.
VETICAL GARDEN 난간을 활용하여 화분으로 이용 효율을 높임
PLAYGROUND 낙후된 놀이터 대신 계단 밑 공간을 활용하여 놀이 공간 제공
DRYING RACK 넓은 계단 참 공간을 이용하여 빨래 공간 제공
BICYCLE RACK 입구의 계단 밑 공간을 이용하여 자전거를 보관할 수 있도록.
COMMUNITY 계단에 앉을 수 있도록 공간을 제공하여 주민 활동 활성화.

Banister that is composed of stairs for sitting is used to increase efficiency of use through flower pot. Stairs and space are used instead of primitive playground to provide space for playing. Wide staircase space is used to provide space for laundry. Entrance stairs and space are used to store bicycle. Space is provided to enable residents to sit on stairs, thus activating resident activity. Slope and stairway are selected by residents. Space under stairs provides playing space for children. Stairs are used as chairs to form community space for residents.

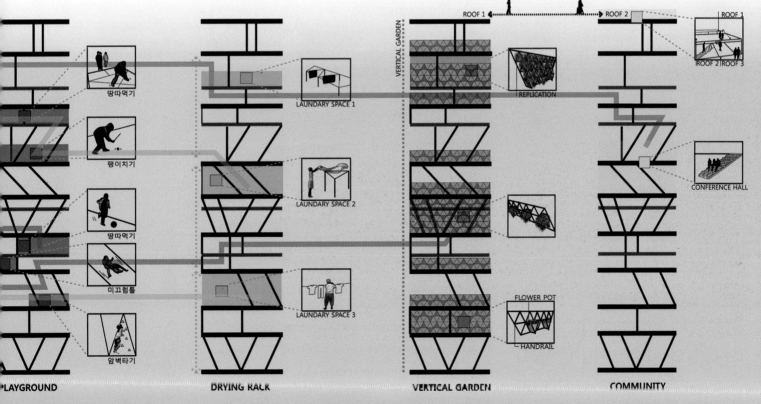

PLAYGROUND   DRYING RACK   VERTICAL GARDEN   COMMUNITY

01 난간에 화분의 역할을 부여하여 미적효과를 준다.

02 다양한 경로를 제공하여 주민들에게 길 선택의 기회를 준다.

03 다양한 경로를 제공하여 날마다 새로운 경로로 다닐 수 있다.

01 Role of flower pot is provided to banister, giving aesthetic effects.

02 Various routes are provided to allow residents to select passage.

03 Various routes are provided to enable residents to use new passages everyday.

BICYCLE RACK 2

ROOF 1
ROOF 2 ROOF 3

04 Vertical garden is used to enhance efficiency of space.

05 Abandoned stairs and space are used as bicycle rack.

06 Rooftop space of each building is connected to improve mobility.

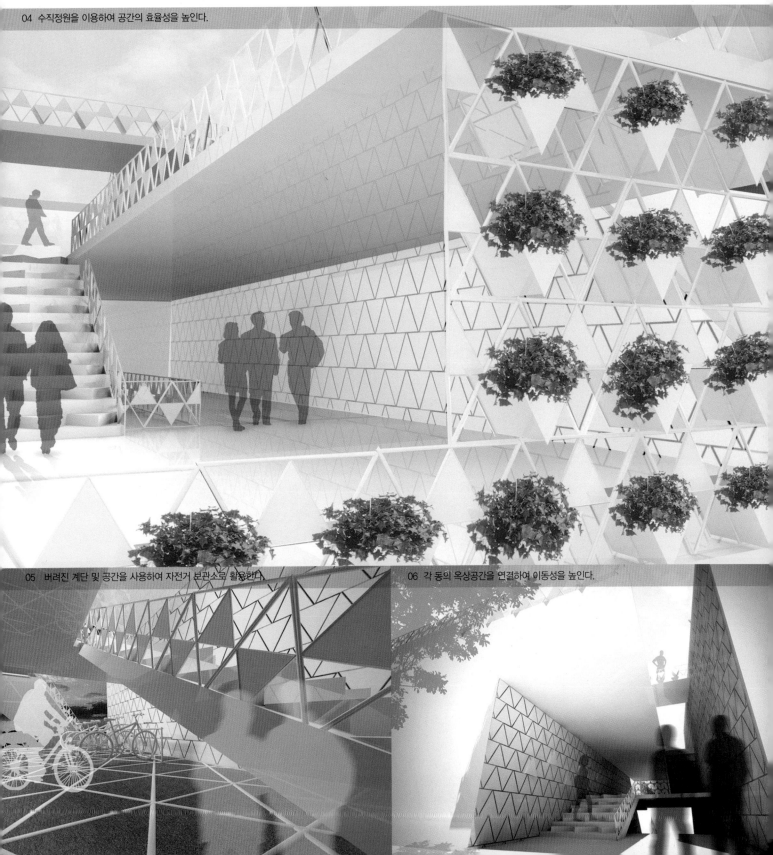

04 수직정원을 이용하여 공간의 효율성을 높인다.

05 버려진 계단 및 공간을 사용하여 자전거 보관소로 활용한다.

06 각 동의 옥상공간을 연결하여 이동성을 높인다.

문래동 철재 공장은 숨쉬지 못하는 곳이였다. 하지만 나비 5마리 가 날아와 문래동 철재공장 옥상에 앉음으로써 그 곳은 이제 더 이상 숨쉬지 못하는 곳이 아닌 숨쉬는 자연친화적인 공간이 되었다.

The iron plant where you cannot breathe - the iron plant where you can breathe The iron plant was the place you cannot breathe. With 5 butterflies flying to the top of the plant, it came to the eco-friendly place you can breathe.

나비,
문래동에 날아 오르다

Butterflies fly in the Munrae-dong

It is said that butterflies fly only on bright days. With the wish for future bright days of Munrae-dong, I made 5 butterflies on the top of building in Munrae-dong.

SITE

서울 특별시 영등포구 문래동 37번지 문래동 철재상가 단지
Munrae-dong iron shopping center in #37
Munrae-dong Youndungpo-gu Seoul

FEATURE

80년대 초 철재상가단지라는 명성을 얻었지만, 시대가 변해감에 따라 수도권 주변의 땅값이올라 높은 땅값을 견디지 못한 사람들이 하나 둘씩 외각으로 공장을 옮기기 시작하면서 현재는 철재상가 단지라는 명맥만 유지하고 있다. 80년대 초 철재상가단지라는 명성을 얻었지만, 시대가 변해감에 따라 수도권 주변의 땅값이 올라 높은 땅값을 견디지 못한 사람들이 하나 둘씩 외각으로 공장을 옮기기 시작하면서 현재는 철재상가 단지라는 명맥만 유지하고 있다.

While Munrae-dong iron shopping center had its name as a iron shoppincenter during early 80', it is currently barely surviving because shop owners have moved to outskirt of Seoul with the rise of cost of land.

SITE ANALYSIS

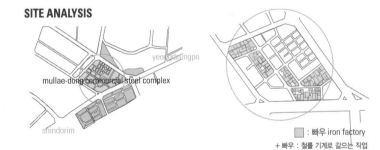

FLOOR

Each plant has 1st and 2nd floor.

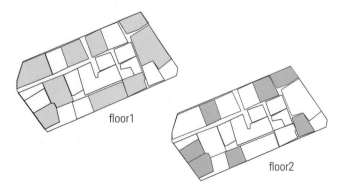

floor1

floor2

SITE PROBLEM

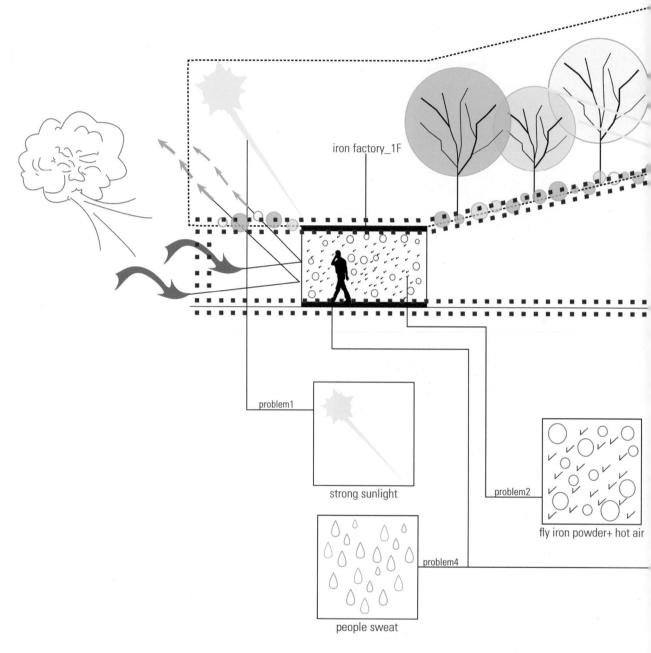

iron factory_1F

problem1

strong sunlight

problem2

fly iron powder+ hot air

problem4

people sweat

DESIGN BASE

ventilation(환기)
circulation(순환)
환기 : 탁한 공기를 맑은 공기로 바꿈
순환 : 주기적으로 자꾸 되풀이하여 돎

Ventilation : to allow fresh air to enter and move around a room, building, etc.
Circulation : the movement of something around an area or inside a system or machine.

DESIGN MOTIVE

pipe : 물이나 공기, 가스 따위를 수송하는 데 쓰는 관
butterfly : 낮에 활동하는 무리를 통틀어 이르는 말몸은 가늘고 빛깔이 매우 아름다움 머리에 한 쌍의 더듬이와 두 개의 겹눈이 있고 가슴에 큰 잎 모양의 두 쌍의 날개가 있음

pipe : a tube through which liquids and gases can flow.
butterfly : a flying insect with a long thin body and four large, usually brightly coloured, wings. it has a pair of fellers and compound eyes on the head and big leaf shaped two wings one the breast.

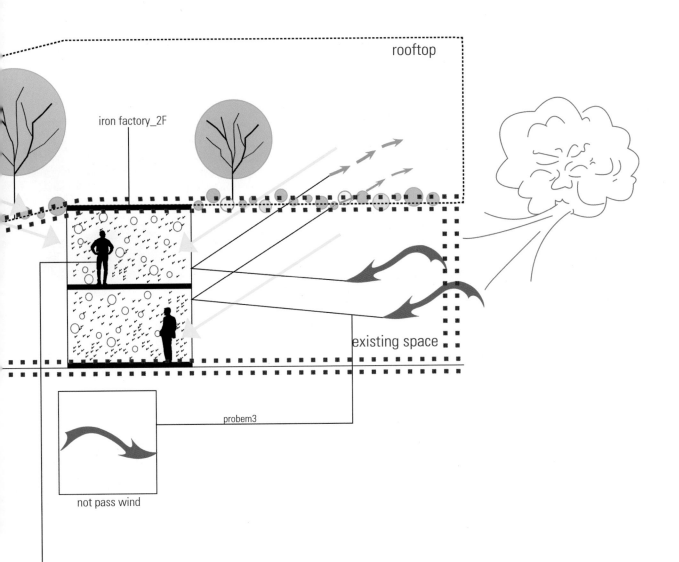

rooftop

iron factory_2F

existing space

probem3

not pass wind

This site has 4 problems. First one is strong sunshine. Hot sunlight comes directly to the plant because of not having sunlight protection which shade the plant. Second one is raising iron powder and hot air. Third one is the wind which cannot go through the plant. The wind cannot go through the inside of the plant causes the fourth problem, sweating workers. Also the workers of this site are the Pau, the ones who shavethe iron. So the raising of iron powder also cause a problem.

PURPOSE

문래동 철재상가는 파이프를 주로 취급하는 곳이다. 또한 파이프는 다양한 각도로 꺾을 수 있을 뿐만 아니라 문래동 철재상가의 특징을 잘 보여줄 것이다. 또한 나비는 맑은날에만 나는 곤충이다. 앞으로 문래동의 맑은 앞날을 기대하며 빠우공장옥상에 나비 5마리를 날아오게 해주려고 한다.

Munrae-dong iron shopping center deals mainly with pipes. The pipe can be bended with various angles and it will be able to show the characteristics of the center. Also the butterfly only flies on the clear days. I want 5 butterflies to fly on the top of plant Pau with the wish for future bright days of Munrae-dong.

The pipes have 4 kinds. First pipe is air inject one and its thickness is 2000. Second one is iron powder + water pipe and its thickness is 900. Third one is sunlight protection pipe and its thickness is 500. Fourth one is water pipe and its thickness is 300.

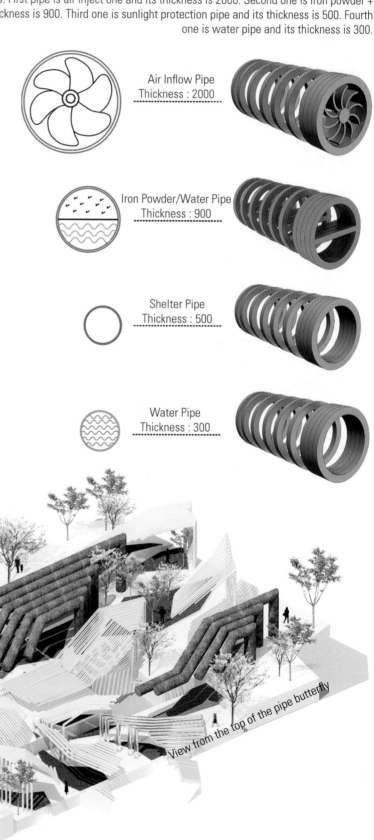

Air Inflow Pipe
Thickness : 2000

Iron Powder/Water Pipe
Thickness : 900

Shelter Pipe
Thickness : 500

Water Pipe
Thickness : 300

View from the top of the pipe butterfly

Plant butterfly people in the space of a pipe break

PROBLEM

To solve the first problem of strong sunlight, the protection with the wings of the butterfly is used. The protection pipe among 4 kinds pipes is the 500-thickness one and it makes shade where people can take a rest. The second problem of raising iron powder can be the food for phytoplankton and hot air can be solved with using vaporization. The iron powder and water pipe is 900-thickness and iron powder can be food for phytoplankton by going into the water while being injected though half-divided pipe. Also phytoplankton can be helpful to global warming with the absorption of CO2. The water going through the pipe solves theproblem of hot air with the vaporization of around heat. The third problem of blocking of wind can be solved by air inject pipe. The air inject pipe of 2000-thickness has a propeller inside. So it can inject fresh air into the plant and removes hot air by making aircirculation. The fourth problem that workers are sweating canbe solved by the water pipe. With the waterpipe, workers can remove the sweat by taking shower with the water cominginto the shower booth.

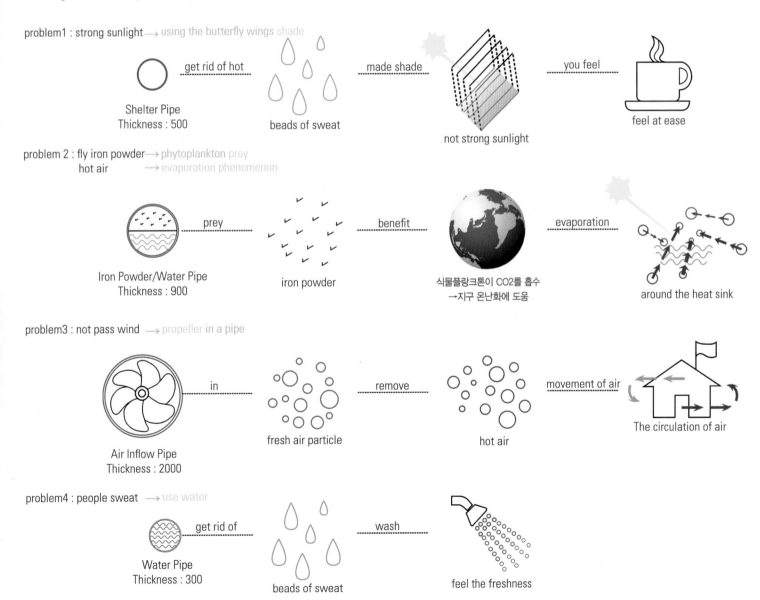

problem1 : strong sunlight ⟶ using the butterfly wings shade

Shelter Pipe
Thickness : 500 — get rid of hot — beads of sweat — made shade — not strong sunlight — you feel — feel at ease

problem 2 : fly iron powder ⟶ phytoplankton prey
hot air ⟶ evaporation phenomenon

Iron Powder/Water Pipe
Thickness : 900 — prey — iron powder — benefit — 식물플랑크톤이 CO2를 흡수 →지구 온난화에 도움 — evaporation — around the heat sink

problem3 : not pass wind ⟶ propeller in a pipe

Air Inflow Pipe
Thickness : 2000 — in — fresh air particle — remove — hot air — movement of air — The circulation of air

problem4 : people sweat ⟶ use water

Water Pipe
Thickness : 300 — get rid of — beads of sweat — wash — feel the freshness

DESIGN PROCESS1

1 pattern extraction   **2** the movement of pattern

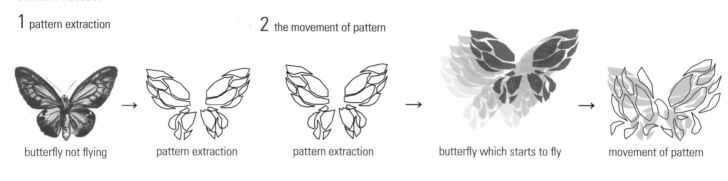

butterfly not flying   pattern extraction   pattern extraction   butterfly which starts to fly   movement of pattern

SPACE ZONING

There are 3 themed chairs and a shower booth in the resting place.

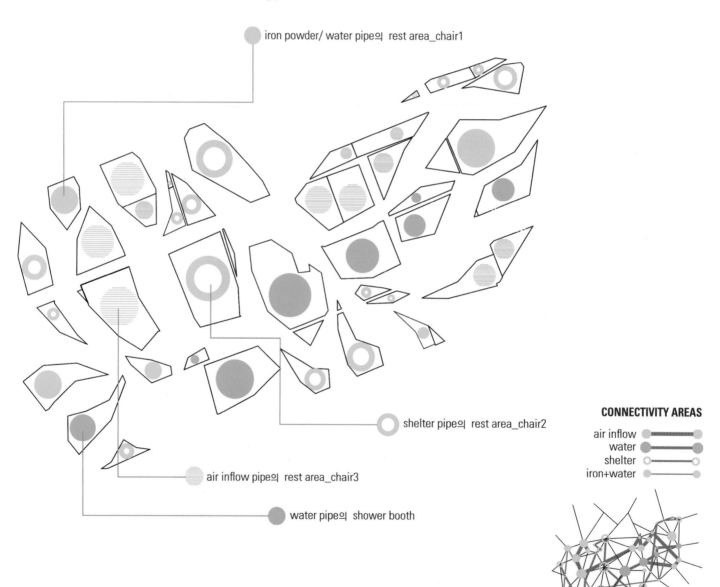

iron powder/ water pipe의 rest area_chair1

shelter pipe의 rest area_chair2

air inflow pipe의 rest area_chair3

water pipe의 shower booth

CONNECTIVITY AREAS

air inflow
water
shelter
iron+water

TRAFFIC LINE

▶▶▶▶▶▶▶▶▶▶ route1
▶▶▶▶▶▶▶▶▶▶ route2
▶▶▶▶▶▶▶▶▶▶ route3

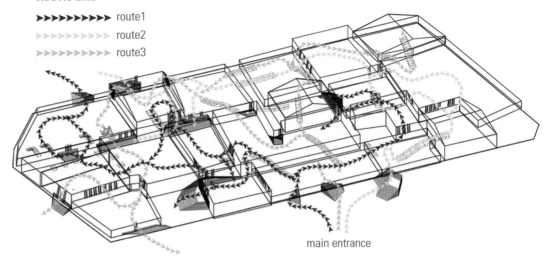

main entrance

DESIGN PROCESS2

3 pipe vending

regular pipe → getting lines from the shape of butterfly → vending the pipe with following the lines + → vended pipe following the shape of butterfl

Pipes Butterfly in the sky like flies

Iron powder/ Water pipe의 Rest Area

CHAIR1

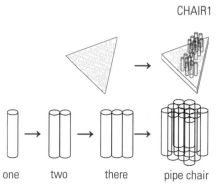

one → two → there → pipe chair

Shelter pipe의 Rest Area

CHAIR2

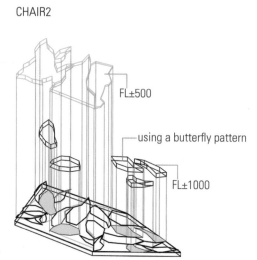

FL±500

using a butterfly pattern

FL±1000

CHAIR3

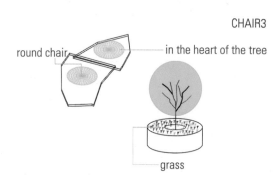

round chair

in the heart of the tree

grass

Air inflow pipe의 Rest Area

EXPLODED VIEW

① the main role of ventilationand circulation
② shelter and clean and refreshing shower
③ eco-friendly materials
④ dirty and contaminated factories

Stone, wood, grass

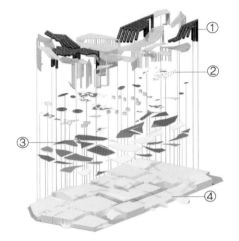

The place where the butterfly can breathe

the iron plant where you cannot breathe

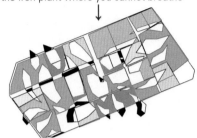

the iron plant where you can breathe

SHOWER BOOTH

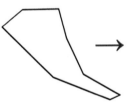

pattern of a butterfly

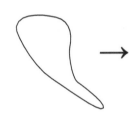

creating the pattern shape

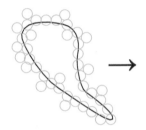

roll along the lines of pipe

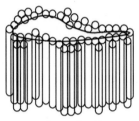

shower booth

Water pipe - Shower Booth

EPOCHE

Moonlae-dong 1-ga

에포케, 모든 판단을 중지하고 관람하라

대상지 주변은 상대적으로 빠르게 발전하고 있다. 영등포와 신도림 일대는 서울 서남권의 영향력있는 부도심 으로 성장했다. 주위에는 고업무용 고층빌딩과 아파트가 건설되었으며 서울과 경기도 일대를 연결하는 교통이 발달되어 있다. 새로 개발된 주거지역에는 구매력 높은 인구가 유입되면서 고급 상업시설과 학교 등이 신설되었다. 그러나 문래동은 철길과 주거지의 완충 역할을 하고 있기 때문에 과거의 모습이 남게 되었다. 많은 발전에도 불구하고 지역적 특성을 담고 있는 문화공간은 부족한 현실이다. 따라서 기존 철공소 지붕의 높낮이를 이용하여 예술인을 위한 '창작공간'을 구축한다. 그럼으로써 문래동과 지역의 발전에 중요한 역할을 하게 된다.

The peripheral aboveground areas are growing at a relatively fast pace. Yeongdeungpo and Sindorim areas have grown to be a subcenter with strong influence in the southeastern part of Seoul. As people with high-purchasing power came into newly developed residential areas, high-class commercial facilities and schools were built. But as Moon-rae-dong acts as a buffer between the railroad and the residential area, it still has traces of its past left to a substantial degree. Despite undergoing a lot of growth, cultural spaces with regional characteristics are lacking. Therefore a "creative space" for artists is constructed using the varying heights of the roof of small-sized steel plants. It is expected to play an important role for the growth of Moon-rae-dong and the surrounding area.

SITE ANALYSIS

Moonlai-dong 1-ga Youngdeungpo-gu

Ironworks

High-rise residential

Crossroads

Ground line

Commercial space

Schools

대상지는 과거 구로공단의 소규모 철공소 밀집지역으로 최근 주변의 개발로 인해 고립된 섬 형태로 남게 된 지역이다. 최근 산업의 패러다임의 변화가 소규모 철공 수요의 변화를 가져왔고 임대 수요가 줄어드는 결과를 초래했다. 그러나 저렴한 임대료와 자유로운 작품 공간을 원하는 예술인들이 그 수요를 대신하면서 조금씩 지역이 되살아나고 있다. 버려진 공간에 예술가들이 모여든 건 5~6년 전이다. 현재는 50개 정도의 창작실과 전시 공간 입주해 있다. 활동하는 예술가들의 숫자도 130명에 이른다. 활동하고 있는 예술 장르 또한 다양해서 퍼포먼스, 악기, 문화 활동가 등이 그룹을 형성하고 있다. 그야말로 '창작촌'이라는 말이 어울리는 곳으로 문래동이 변화한 것이다.

The project site is an area filled with small-sized steel plants of the former Guro Industrial Complex. Due to recent developments of the surrounding area, it is now left isolated in the shape of an island. Changes in the industrial paradigm brought changes to the demands for the small-sized steel plants, which resulted in the reduction in the demand for leasing. But the gap in the demand has since been met by artists who wanted cheap lease prices and flexible space in which to display their works. This is reviving the area little by little.

It was five to six years ago that artists gathered in the abandoned area. Currently there are about 50 places occupied for creative work and exhibition space purposes. There are as many as 130 active artists, whose various genres include performance, instrument, and culture. Moon-rae-dong has undergone change to a place befitting its label, "Creative Community."

대상지 주변은 상대적으로 빠르게 발전하고 있다. 영등포와 신도림 일대는 서울 서남권의 영향력있는 부도심으로 성장했다. 주위에는 고업무용 고층빌딩과 아파트가 건설되었으며 서울과 경기도 일대를 연결하는 교통이 발달되어 있다. 새로 개발된 주거지역에는 구매력 높은 인구가 유입되면서 고급 상업시설과 학교 등이 신설되었다. 그러나 문래동은 철길과 주거지의 완충 역할을 하고 있기 때문에 과거의 모습이 남게 되었다.

The peripheral aboveground areas are growing at a relatively fast pace. Yeongdeungpo and Sindorim areas have grown to be a subcenter with strong influence in the southeastern part of Seoul. As people with high-purchasing power came into newly developed residential areas,

high-class commercial facilities and schools were built. But as Moon-rae-dong acts as a buffer between the railroad and the residential area, it still has traces of its past left to a substantial degree.

Ironworks underserved areas Moonlai-dong

Internal Construction Ironworks

많은 발전에도 불구하고 지역적 특성을 담고 있는 문화공간
은 부족한 현실이다. 따라서 기존 철공소 지붕의 높낮이를 이
용하여 예술인을 위한 '창작공간'을 구축한다. 그럼으로써
문래동과 지역의 발전에 중요한 역할을 하게 된다. 옥상에 구
축되는 창작촌' 구조물을 지상으로 운행되고 있는 전철 이용
객에게 노출시킴으로서 홍보효과를 얻어낼 수 있다.

Despite undergoing a lot of growth, cultural
spaces with regional characteristics are lacking.
Therefore a "creative space" for artists is constructed using the varying heights of the roof of
small-sized steel plants. It is expected to play an
important role for the growth of Moon-rae-dong
and the surrounding area. The creative space to
be built on the roof would be able to be shown to
passengers of aboveground subway, with promo-
tional benefits expected.

낙후된 철공소에는 인적이 드물지 않게 되었고 주위는 신도
림, 영등포로 인해 인구이동이 많은 곳 사이에 자리 잡은 문래
동, 사람들에게 산업화의 잔재라는 인식에서 벗어나 예술인
으로 하여금 철공소에 사람의 이목을 이끌려고 한다.

As the remnant of industrialization, the region
lagged behind, assuming noise, different roof
levels, ground subway trains, air pollution and
removal of high roads. The deserted iron works
have become quite busy places as Moonlai-dong
is located in a place where large population passes by due to adjacent Shindorim and Yongdeung-
po. Breaking away from the notorious recognition
as the remnant of industrialization, artists draw
public attention to iron works.

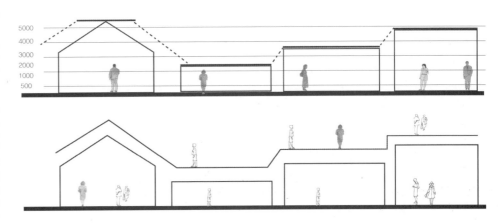

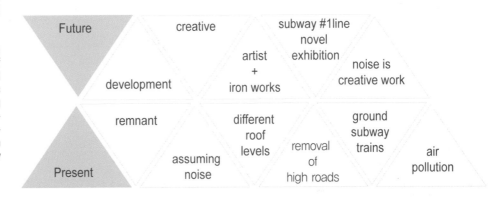

abandoned spaces

old facilities

CONCEPT PROCESS

Rubik's Twist

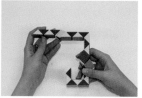

긴 막대기형태 루빅스 트위스트는 동일한 크기의 삼각 모듈이 조합된 어린이 장남감이다. 누구든지 모듈의 연결부위를 돌려서 임의의 형태를 만들어 낼 수 있다. 여러 가지 형태조합이 가능하고 일정한 답을 찾는 과정이 아닌 창의적으로 형태를 민들어 가는 것으로 흥미를 일으키는 써즐이다. 부빅스 트위스트의 형태를 적용한 창작 예술품 전시공간을 계획한다. 근대이후에 일반화된 합리성에 근거한 입방체적 전시공간에서 탈피한다. 언제나 동일한 프로그램이 전시되는 공간을 거부한다. 루빅스의 특징을 적용하여 공간 내의 상·하·좌·우에 대한 규칙을 무시하고 전시 프로그램에 따라 언제나 새로운 공간을 형성하도록 계획한다.

Rubik's Twist is a children's toy in which same-sized triangular modules are assembled. It is in the shape of a long stick. Anyone can create an arbitrary shape by rotating the connections of modules and many combinations are possible. It is a puzzle that is interesting not because there is a single answer to which you try to reach but because you get to draw on your creative side to make various original shapes. A creative art exhibition space in which Rubik's Twist's shape is used is planned. It breaks away from the exhibition space in the form of a cube based on rationality which has come to become the standard since the modern times. Rubik Twist's characteristics are applied to ignore the rules regarding spatial top/bottom/left/right and to always create a new space according to the particular exhibition program.

KEYWORD

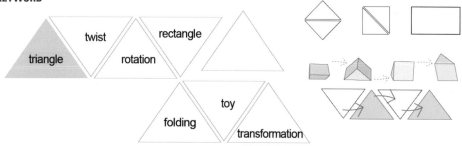

triangle · twist · rectangle · rotation · folding · toy · transformation

루빅스의 회전은 공간의 회전을 보여주며 루빅스 하나하나의 삼각형들이 모여 공간을 형성하도록 계획한다.

Rotation of a Rubik's Cube shows a spatial rotation, and each triangle of the Rubik's Cube gathers to combine the space.

Morning

Nightfall

Nightview

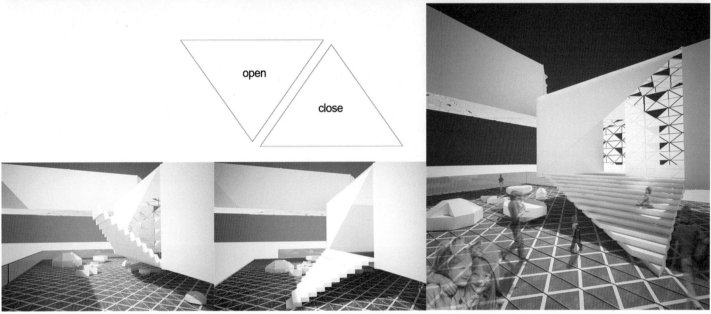

open

close

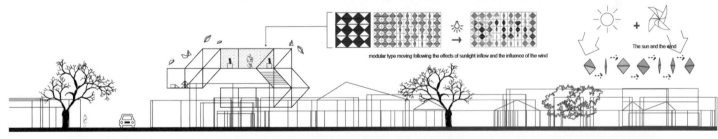

modular type moving following the effects of sunlight inflow and the influence of the wind.

The sun and the wind

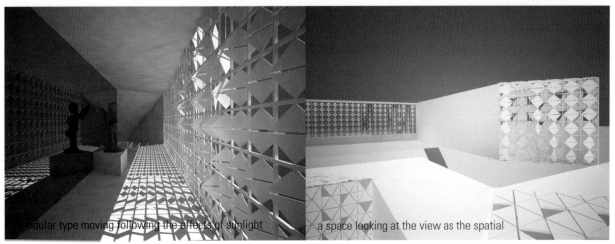

modular type moving following the effects of sunlight

a space looking at the view as the spatial

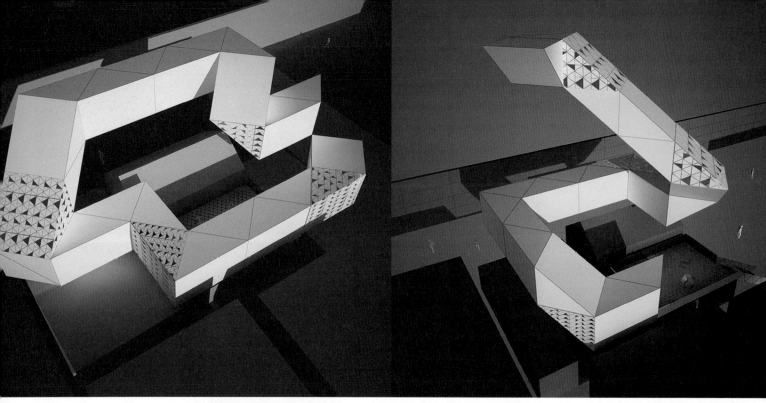

JANUARY / FEBRUARY / MARCH

APRIL / MAY / JUNE

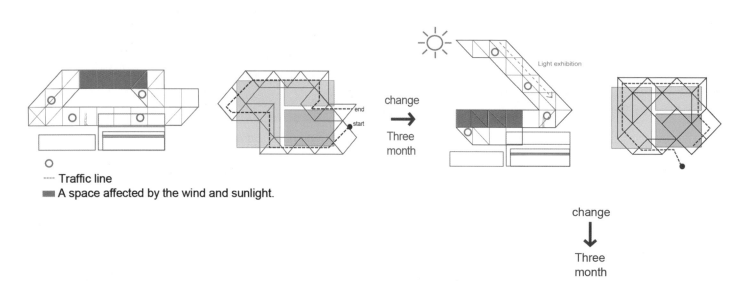

Light exhibition

change

→

Three
month

change

↓

Three
month

change

←

Three
month

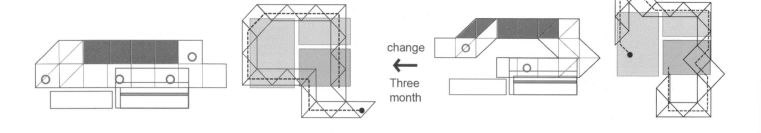

○

---- Traffic line

■ A space affected by the wind and sunlight.

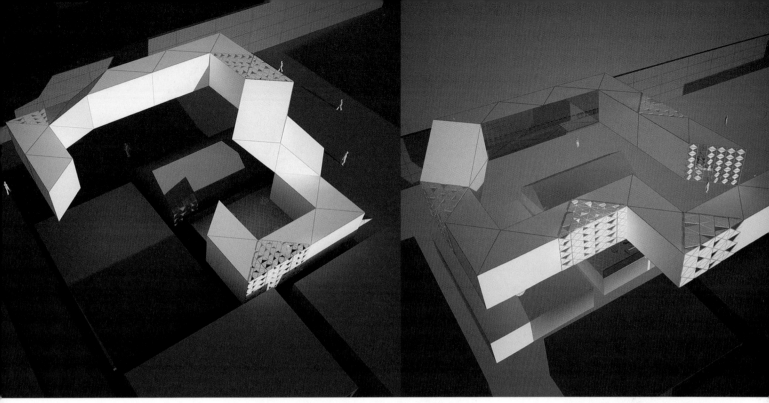

JULY / AUGUST / SEPTEMBER   OCTOBER / NOVEMBER / DECEMBER

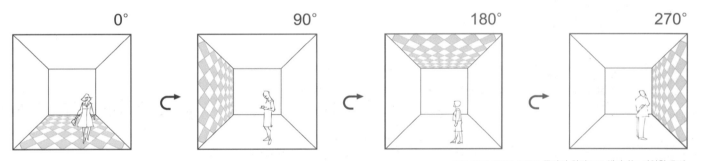

0°   90°   180°   270°

0도 90도 180도 270도 공간의 회전으로 생겨나는 다양한 효과
Diverse effects arise from the spatial rotation at 0, 90, 180 and 270 degrees.

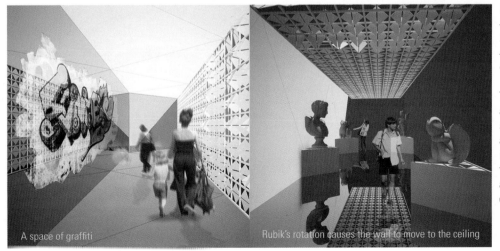

A space of graffiti

Rubik's rotation causes the wall to move to the ceiling

계간 또는 년간 전시계획에 따라 형태를 바꾸는 전시공간은 입주하고 있는 예술인들뿐만 아니라 방문하는 관람객에게도 끊임없는 변화를 감지하게 하는 흥미로운 공간이 되도록 한다. 루빅스의 삼각형 모듈들은 다양한 형태로 조합되어 공간을 형성한다.

The exhibition space would take different shape according to the quarterly or yearly exhibition schedule. It would be interesting not only to the artists residing but to tourists as well, as a space in which they can constantly sense change. The triangular modules of Rubik's Twist would get assembled to create various shapes and in turn creating the space.

루빅스의 특징을 적용하여 공간 내의 상·하·좌·우에 대한
규칙을 무시하고 전시 프로그램에 따라 언제나 새로운 공간을 형성하도록 계획한다.

Rubik Twist's characteristics are applied to ignore the rules regarding
spatial top/bottom/left/right and to always create a new space according to the particular exhibition program.

A space looking at the view as the spatial end

the furniture using the corner points out directionality

VIDEO ART ▶

◀ SCULPTURE

a state when the wall is used as a light

external resting place

stairs

the space of light when a steep space generates

a space of video art

Mark In Town
가리봉동에 흔적 남기기
Leave a trace in Garibong-dong.
There are some countrymen living in limited space in Garibong-dong, missing hometown. I would like to make a community only for them in rooftop to provide an opened space to them.

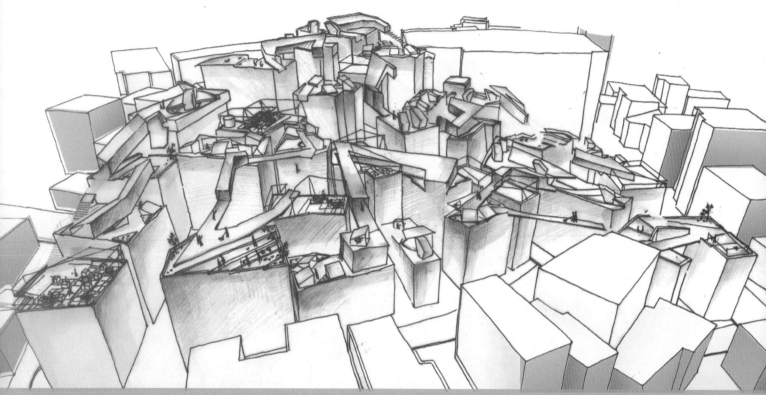

BACKGROUND

일반적으로 사회적 약자라 함은 대부분 노인 아동 여성 등을 생각한다. 하지만 1평 남짓 좁은 공간에서 인간의 기본적인 권리조차 누리지 못하는 공간적 약자도 사회적 약자라 할 수 있다. 가리봉 쪽방촌은 그러한 사회적 약자가 살고 있는 곳이다. 우리는 가리봉동 쪽방촌에 살고 있는 중국 동포들에게 좁은 쪽방 외에 열린 공간을 제공하려한다. 그 열린 공간은 커뮤니티 공간이 되며 일상적인 생활을 하지 못하는 이들에게 기본적인 삶의 권리를 제공해 줄 것이다.

Usually we think of oldmen , kids, or women for social disadvantaged people, however, space-disadvantaged people living in a narrow space - 1 square meter or so and foreign workers living far away from home without any family can be called as a social disadvantaged people.We try to provide an opened space to our fellows, Korean- Chineses living in dosshouse, Garibong-dong. The opened space would be a place for their community and offer the basic right of life to them who do not live quotidian general life.

SITE

서울 구로구 가리봉동
Garibong-dong,Guro-gu,Seoul
"연변타운"
'Yanji town'

60년대 우리나라 제조업 핵심공장이었던 가리봉동 구로공단 1990년 후반부터 서울디지털산업단지로 바뀜, 수많은 공장이전, 노동자들이 떠남, 중국 동포들이 모여 "연변타운" 형성

"쪽방촌"
방하나를 여러 방으로 나뉜 한 평 남짓한 공간.
주택한 채에 32세대가 거주할 정도로 고밀도 지역.

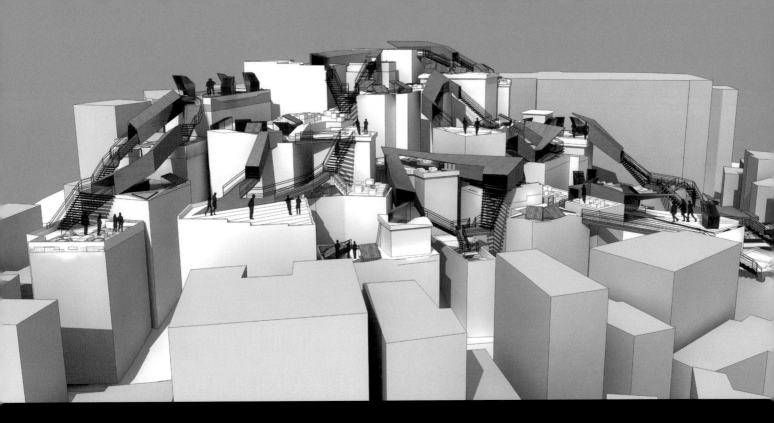

CONCEPT
마작 麻雀을 두디

빡빡한 일정으로 살아가는 조선족들에게 같은 처지의 조선족끼리 서로를 이해하고 커뮤니티를 형성하는 것을 돕기 위한 것으로 마작을 뽑았다. 지금은 가리봉동에 살고 있을지라도 그들은 분명 중국에서 친구들과 마작을 하며 여유를 즐겼을 것이다. 제2의 고향인 가리봉동에 마작을 통해 옛기억을 회상하며 그 곳에 자국을 남긴다. 각 옥상의 용도는 놀이공간 인터넷 전화공간 휴식공간이고, 옥상과 옥상사이를 연결시키는 다리는 중국동포들에게 고향과 현지를 이어주는 매개체가 된다.

Mah-jong was selected to help Korean- Chineses living in a tightened schedule and understanding each other to make a community. Now though they live in Garibong-dong, they may enjoy spare time ,playing mah-jong. Leave a trace in the 2nd hometown, Garibong-dong through Mah-jong , recollecting memories in the past time. Each roof-top would be places for amusement, place for internet &telephone and place to relax. Moreover, the bridge to connect each roof-top would be a medium to connect hometown and local place.

옛기억 → 자국내기
MARK – 고향에 대한 기억 새기고, 다리를 통해 서로 연결을 해준다.

룰 → 반시계방향
ROTATE – 전화공간에 반시계방향으로 도는 게임 룰을 적용시킨다.

패 → 모듈
MODULE – 웹서핑개인실 전화실 , 마작패를 모듈로 삼는다.

조 → 결합
COMBINATION – 조를 먼저 만들면 이기는 게임으로 조를 만드는 것은 결합이라 할 수 있다.

SPACE ANALYSIS

가리봉 PROBLEM	중국 CUSTOM
생활공간 1평	정방형 주거 형태
외로움	위계적 구성
주거시설 노후 +	
공원 등 시설 부족	숫자 8 선호

→

IMPROVEMENT

옥상 연결 → 이동통로

사회적, 문화적 약자의 소통

문화, 휴식 공간
가족과 소통 가능하게 할
인터넷, 전화공간

각 기능의 동 – 8 개구성

다리 – 정방형

→

NECESSARY SPACE

공동체형성, 놀이공간

고향 소식을 접하거나 세상얘기를
알 수 있는 웹서핑 공간

가족과 소통하고, 그리움을 푸는 공간

기본적인 생활 충족시킬 공간

BRDIGE

다리는 조선족에게
고향과 현지를 이어주는 매개체

+

이웃끼리 이어주는 매개체
The medium to connect hometown and local place

+

the medium to connect neighbors

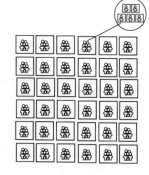
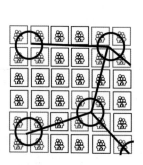

PROCESS

조선족이 많이
살고 있는 연변

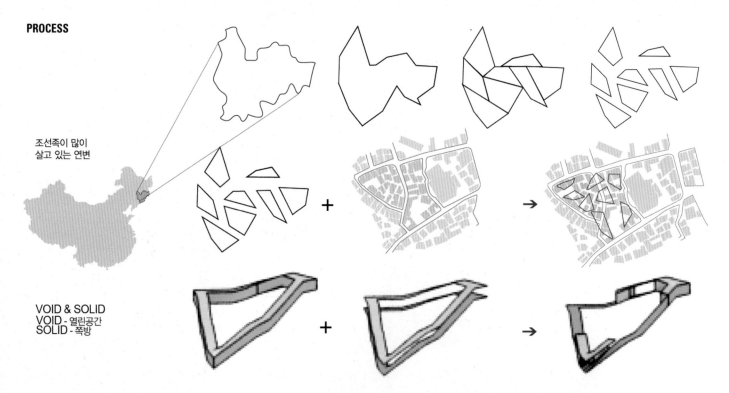

VOID & SOLID
VOID - 열린공간
SOLID - 쪽방

MARK

"단 차이를 이용해 자국을 나타냄"
중국 동포들이 살고 있는 가리봉동에 옛 생각을 떠올리게 하기
위해 '레벨 차' 라는 용어로 자국을 표현 하였다. 가리봉동 특성
상 존재하는 건물의 레벨 차, 각 공간의 특성에 맞게 레벨 차를 적
용시켰다. 자국은 "중국에서의 생활 모습을 현지에 새긴다." 라
는 의미이다.

"The medium to connect hometown and local place,
the medium to connect neighbors."
To think of the memories in the past in Garibong-dong
where Korean- Chineses are living , we expressed
the trace with the terms of 'difference of level'. The
difference of level in the buildings existing by one of
characteristic of Garibong-dong, 'the difference of
level' has been applied to fit for the characteristics
of each space. The trace means " Engrave the scenes
from daily life to the local places. "

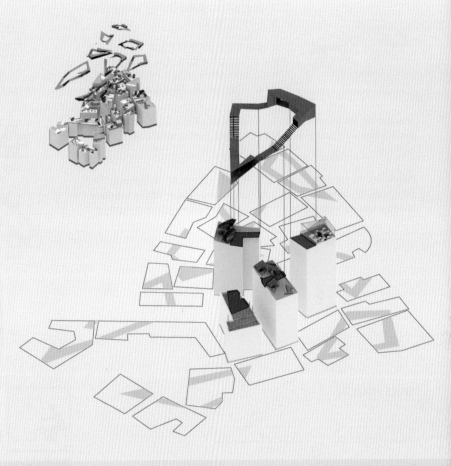

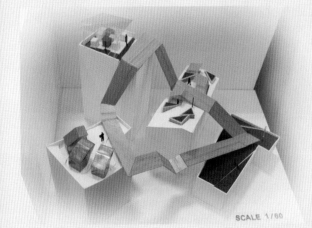

SCALE 1/60

다리를 통해 다른 옥상으로 옮겨 갈 수 있고, 다리 위에서 쉴 수 있다.

PLAYING ZONE

가리봉동에는 중국동포들을 위한 놀이공간이 없다. 그래서 옥상공간에 중국의 대표적인 놀이인 마작을 할 수 있는 공간을 만든다.

마작패를 모듈로 삼아 판을 만든다. 그 판은 의자가 될 수도 있고, 탁자가 되기도 하며, 이동통로가 되어 녹지로 쓰인다. 햇빛을 차단하기 위해 모듈의 격자 그리드에서 따온 그늘막을 설치 하였다.

There is no amusement place for Korean Chineses in Garibong-dong, so we will make a space to play Mah-jong, the Chinese traditional game in roof-top.

Regarding the tile of Mah-jong as a module, a board would be made. The board may be used for a chair, table or passage. To cut off the sunshine, shadow tent copying the grid shape from the module would be built up.

SPACE PROCESS

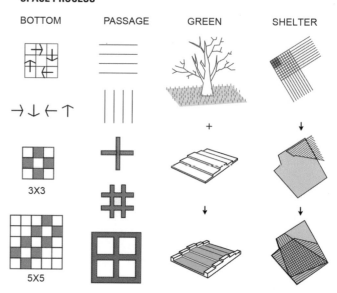

SPACE ANALYSIS

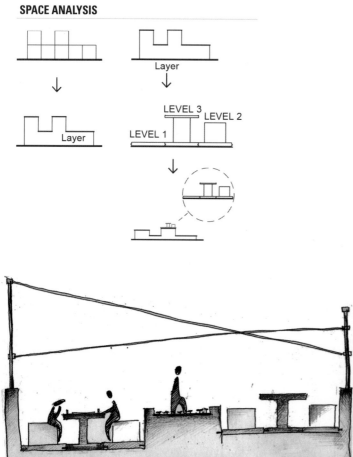

이동부분에 레벨차를 주어 녹지로 활용.
의자가 움직일 수 있어 이동통로가 유동적이다.

쉘터의 상부와 기둥부분에는 약간의 기울기를 준다.

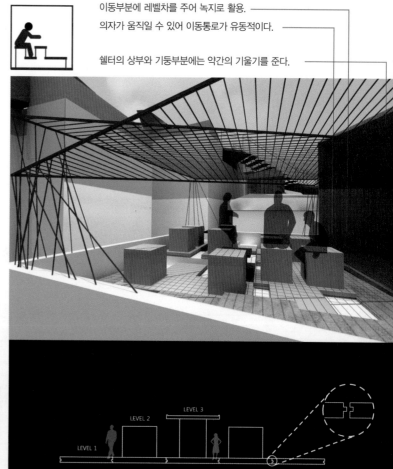

INTERNET ZONE

현재사회는 정보화시대 이지만 가리봉동에는 인터넷을 할 공간이 없다. 그래서 옥상 공간에 고향의 소식을 접할 수 있는 인터넷 공간을 만든다.

사각모듈과 삼각모듈의 결합과 그 결합의 이어지는 라인을 바닥으로 내려 바닥 경사를 주었다. 각 모듈의 형태는 사각형과 삼각형이 슬라이스된 변형된 형태이다. 레벨 차이로 생긴부분으로 인해 컴퓨터실 안으로 들어갈 때는 계단을 만들어 주었다.

Modern society is in information age, however, there is no space to do internet in Garibong-dong. So the place for internet would be installed to hear the news of hometown.

With combination of square module and triangle module , we made a slope ,falling down the lines connected from the combination to the floor. The shape of each module is a transformed shape sliced fromsquare and triangle. The parts made from the difference of the levels ormed steps to enter in the computer room.

SPACE PROCESS

MAS PROCESS

GREEN

FURNITURE

mass   mass

down

cut   cut

PLAN   ELEVATION

SPACE ANALYSIS

계단을 통해 컴퓨터하는 부스로 들어갈 수 있다.
컴퓨터 부스 주변에 녹지와 경사를 줌으로써 지루함을 피할 수 있다.

TELE ZONE

가리봉동 시장에 전화방이 있지만 너무나 비좁다. 그래서 옥상공간에 그들의 외로움을 채워주고, 그리움을 덜어줄 전화 통화하는 공간을 만든다.

전화공간에 반시계방향으로 도는 마작의 룰을 적용시킨다. 전화 부스는 사각형을 밀고 쪼갠 형태이며, 전화 부스 안에는 의자 또는 메모할 종이 등이 있다. 전화공간의 평면은 반시계 회전 방향에 따라단계별로 레벨이 다르다.

In Garibong-dong, there is a telephone room in the market. So the place for telephone would be installed to fill up the lonliness and get over missing. Apply the rule of Mah-jong turning in counterclock wise to the space for telephone.

The telephone booth is a shape pushing and splitting the square. here is a chair and paper to take a note in the booth. The flat surf-ace for space of telephone has different levels at each step by ro-tation direction.

SPACE PROCESS

levelplan

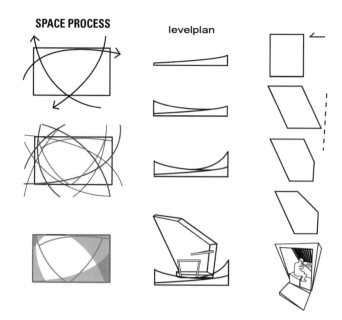

SPACE ANALYSIS

STEP.1

STEP.2

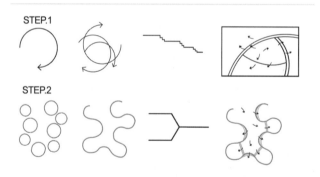

바닥에는 회전 모양의 레벨이 존재한다.

전화부스에는 메모할 수첩 등이 있다.

VOID

가리봉동 쪽방촌엔 공용세탁기는 존재하지만 빨래 널을 공간이 협소하고, 녹지가 현저히 부족하다. 그래서 옥상공간에 빨래를 널고 휴식할 공간을 만든다.

녹지가 있는 계단은 앉아서 쉴 공간이 될 수 있다. 평면상 사선으로 나타내어 지루함을 피할수 있으며, 그 계단은 가구가 될 수 있고, 공간 자체가 될 수도 있다. 건물의 난간은 계단식으로 옥상의 계단을 따라 높이가 다르다.

In dosshouse of Garibong-dong, there are public washing machines, but the space to hang out the wash are lack and the green fields are notably needed. Therefore the place to hang out the wash and take a rest would be built up.

The steps of the green field would be a space to sit down and take a rest. Apparently it looks like a diagonal line and can avoid boredom . Moreover, the steps can be furniture and space itself. The banister of the building has different hight at each step of roof-top .

SPACE PROCESS

STEP. 1

STEP. 2

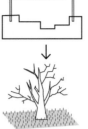

SPACE ANALYSIS

ROOF TOP

빨래를 한다.　　빨래를 널 곳이 없다.　　옥상에 빨래를 넌다.

빨랫줄에 빨래를 널을 수 있게되었다.

계단에 앉아서 쉬거나 이야기를 나눌 수 있다.

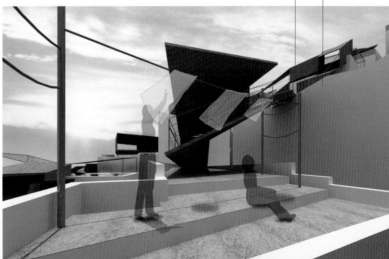

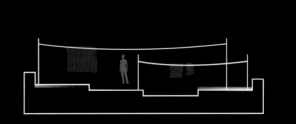

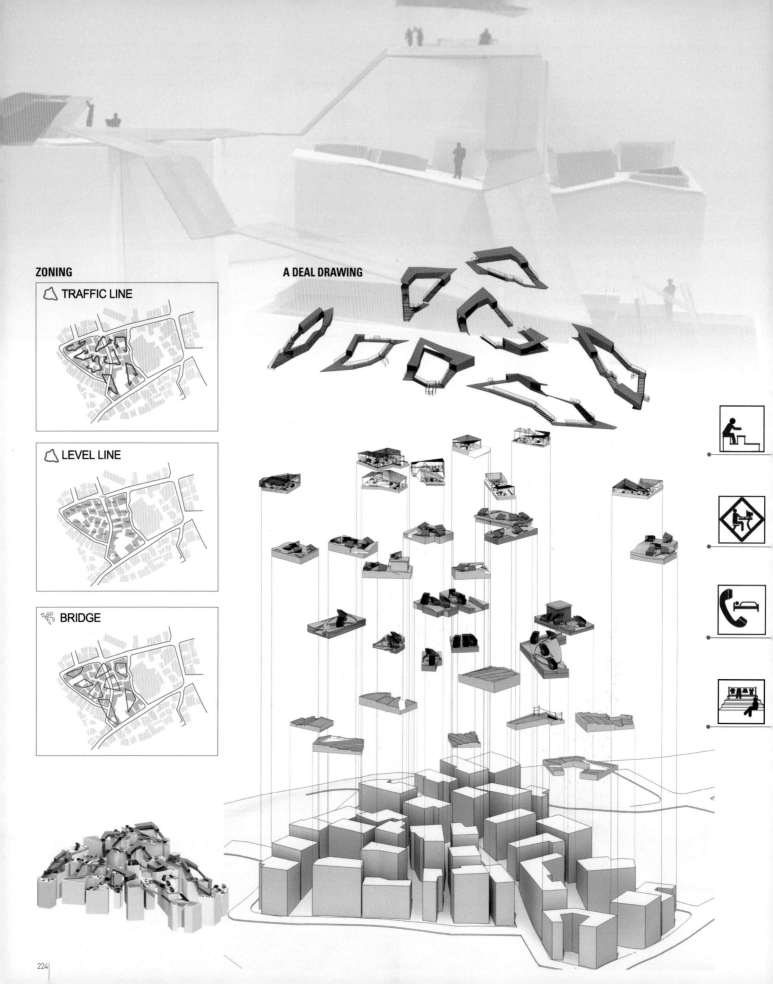

ZONING

TRAFFIC LINE

LEVEL LINE

BRIDGE

A DEAL DRAWING

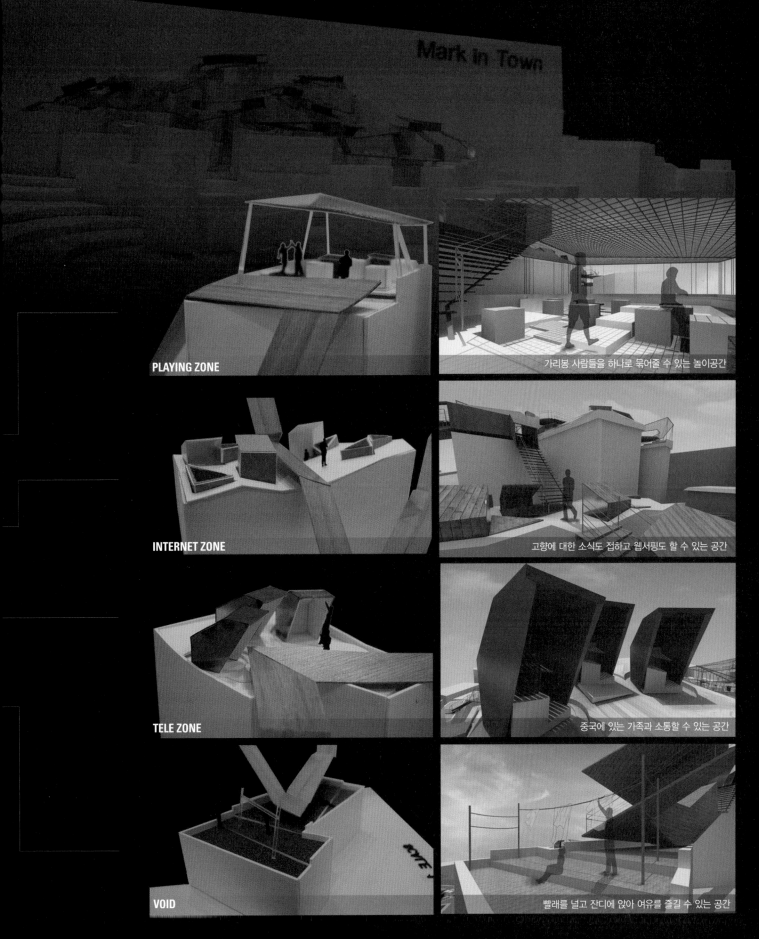

Mark in Town

PLAYING ZONE

가리봉 사람들을 하나로 묶어줄 수 있는 놀이공간

INTERNET ZONE

고향에 대한 소식도 접하고 웹서핑도 할 수 있는 공간

TELE ZONE

중국에 있는 가족과 소통할 수 있는 공간

VOID

빨래를 널고 잔디에 앉아 여유를 즐길 수 있는 공간

SELF - NAISSANCE

스스로 생성하다

상처 입은 자연을 치료함으로써 사람 스스로 치료 받을 수 있는 세운상가 옥상 공간 프로젝트
Sewoon Plaza roof space project for healing wounded nature to treat people

BACKGROUND

인간에 의해 자연이 무분별하게 훼손되고 있다. 쓰레기로 산을 뒤덮고 생활하수로 강과 바다를 더럽히고 화학제품으로 땅도 오염시켰다. 또한 단지 자연 경관을 파괴한다는 이유만으로 멀쩡한 건물을 철거 하고 쓰레기를 만들고 다시 짓기를 반복하고 있다. 자연이 아파하고 상처받았다는 것은 알지만 치료하는 방법을 몰라 소극적으로 행동하는 것은 아닐까? 인간은 상처 입은 몸과 마음을 자연으로부터 치료 받길 원한다.상처 입은 자연을 직접 치료하면서 인간인 우리 또한 자연으로부터 치료받는 공간, 어디서든지,

언제든지, 누구든지 사용가능한 공간이 더 늦기 전에 필요하다.

Nature is being thoughtlessly damaged by people. Mountains are covered with garbage, rivers and oceans are dirtied by sewage and the ground is contaminated with chemical products. Furthermore, buildings in perfect condition are demolished to be made into waste and reconstruction is repeated for the simple reason that it destroys natural scenery. Are humans acting

passively because they are unaware of how to treat nature even though they realize that nature is wounded and hurting? Humans hope that nature will treat their wounded bodies and minds. A place for treating wounded nature and simultaneously facilitating healing from nature, a place that can be used anywhere, anytime, and by anyone is required in today's society before it is too late.

SITE

종로구 종로3가와 퇴계로3가 사이에 있는 우리나라 최초의 주상복합 건물로 북쪽에는 종묘와 남쪽은 청계천이 흐르고 있다. 지하철 1, 2, 3, 5호선이 주변에 위치해 어느 지역으로도 이동하기 편리하다. 본 프로젝트의 대상지 세운상가는 제2차 세계 대전 당시 폭격기에 의한 화재를 막기 위해 조성된 소개 공지에 1968년 김수근 건축가에 의해 설계 되었다. '세상의 기운이 이곳으로 모이라' 라는 뜻으로 세운 (世運) 이라고 지어졌다. 하지만, 1995년 단지 거대한 건물이 도시 자연 경관을 파괴한다는 이유만으로 철거 계획에 들어갔고 국민 소득이 높아지면서 그 자리에 공원을 조성하자는 여론이 형성되었다.

2009년 종로 세운상가를 시작으로 일대는 세운 초록띠 녹색공원으로 재탄생 되었다. 2012년까지는 퇴계로의 모든 상가를 철거할 계획에 있다.

Jongmyo Shrine is located north of the first multipurpose building in Korea positioned between Jongno-3-ga and Toegye-3-ga, while Cheonggye Creek flows in the southern side. It is convenient to move to any region as subway lines 1, 2, 3, 5 are located nearby. Sewoon Plaza, the location of the project, was designed by the architect Soo-keun Kim in 1968 on open area created to prevent fires from the bombers during the 2nd World

War. It was named 'Sewoon' with the meaning 'the gathering of all energies in the world'. However, demolition was scheduled in 1995 for the simple reason that natural urban scenery was destroyed by the massive building, and public opinion was formed in the direction of creating a park at the site, along with increased national income. In 2009, the Jongno area was recreated into Sewoon Green Park, starting with the demolition and reconstruction of Sewoon Plaza. All commercial quarters in Toegye-ro are expected to be demolished by 2012.

MOTIVE

 > > >

자연 경관을 파괴한다는 이유만으로 멀쩡한 세운 상가를 철거하려 한다.

Sewoon Plaza is scheduled to be demolished for the simple reason that it destroys natural scenery, despite being in perfectly good condition.

건축물을 철거하게 되면 어쩔 수 없이 산업폐기물이 발생하고 따라서 자연이 파괴되고 세운상가 상인들 또한 삶의 터전을 잃게 됨.

Demolition of building leads to formation of industrial waste which destroys nature, and merchants of Sewoon Plaza also lose their livelihoods.

세운상가 옥상이 자연을 치료하는 도움이 된다면 철거하지 않아도 됨

Demolition is unnecessary if creation of rooftop on Sewoon Plaza can help treat nature.

자연을 치료하다

Treating nature

CONCEPT

FIRST-AID KIT

상처 입은 자연을 치료하고 그것을 다시 이용하는 옥상공간으로서 '치료' 라는 목적에 맞게 컨셉을 구급상자로 잡았다. 구급상자는 가지고 다니며 어디서든 치료할 수 있고 누구든지, 언제든지 사용가능 하다. 그 안에는 여러 가지 비상약품들이 들어있다. 그렇듯이 자연 치유의 목적으로 우리의 옥상공간에 설치되는 kit계획은 다른 어느 장소에도 쉽게 설치할 수 있다. 설치될 kit는 자연을 치료할 수 있는 다양한 기능들로 채워져 있다.

First-aid kit has been selected as the concept, suitable to the purpose of 'treatment' as the rooftop space is created for treating wounded nature and reusing it. A first-aid kit is carried around to treat anyone, anywhere, anytime. Various emergency medications are contained within the kit. In this sense, the kit plan installed in our rooftop space can also be easily installed in any other places with the purpose of treating nature. The installed kit is filled with various functions for healing nature.

KEYWORD

POP UP · TRANSFORMING · CLEAN · VARIOUS GOODS NEEDED · ANYWHERE · ANYBODY · ANYTIME

SCENARIO 01

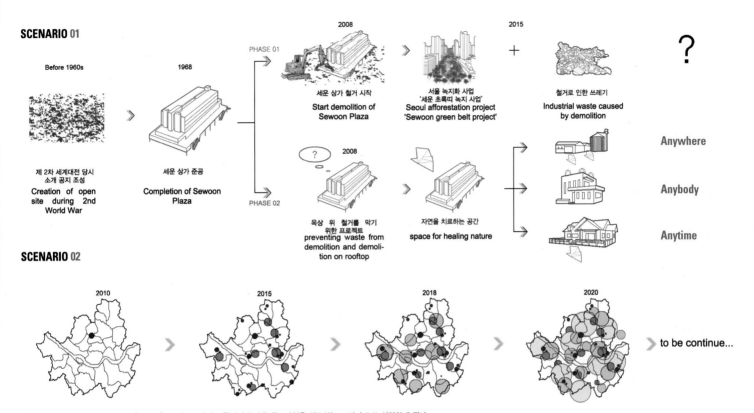

Before 1960s

제 2차 세계대전 당시 소개 공지 조성
Creation of open site during 2nd World War

1968

세운 상가 준공
Completion of Sewoon Plaza

PHASE 01

2008
세운 상가 철거 시작
Start demolition of Sewoon Plaza

2015
서울 녹지화 사업 '세운 초록띠 녹지 사업'
Seoul afforestation project 'Sewoon green belt project'

철거로 인한 쓰레기
Industrial waste caused by demolition

PHASE 02

2008
옥상 위 철거를 막기 위한 프로젝트
preventing waste from demolition and demolition on rooftop

자연을 치료하는 공간
space for healing nature

Anywhere

Anybody

Anytime

SCENARIO 02

2010 > 2015 > 2018 > 2020 > to be continue...

종로구 세운상가에서 시작하여 옥상, 공원, 공장, 등 각지로 확산되어 나중에는 자연을 치료하는 kit의 숲으로 성장하게 된다.
Beginning with Sewoon Plaza in Jongno-gu, it was spread to various places such as rooftops, parks, and factories to later develop into the kit forest for healing nature.

ANYBODY - ANYTIME - ANYWHERE

집 앞 마당에 두고 휴식공간을 이루며 오염된 물을 정화시키고 인분으로 난방과 에너지를 얻으며 태양열 에너지로 전기도 얻을 수 있다.

오염된 물이 많이 나오고 물과 에너지를 많이 필요로 하는 공장 단지에 설치하여 직원들의 휴식공간을 제공하고 에너지를 얻고자 한다.

한강 등 근린공원에 설치하여 시민들이 자연을 치료하는데 사용하고 체험과 동시에 교육효과를 얻을 수 있다.

커다란 상업공간의 옥상 뿐 아니라 개인 주택 등 협소한 장소에도 간단하게 설치하여 옥상에 휴식 공간뿐만 아니라 텃밭을 가꾸기에 좋은 장소가 될 수 있다.

Resting space is formed in the front lawn of house to purify contaminated water and gain heating and energy through excrements and gain electricity through solar heat energy.

It is installed in factory complex where contaminated water in large quantity flows out and large quantity of water and energy is required to provide resting space for employees and gain energy.

It is installed in neighborhood parks such as Han river park to be used for citizens in healing nature, simultaneously gaining experience and educational effects.

It is simply installed not only in rooftops of large commercial space but also in limited space, such as individual homes, to not only be used as resting space on rooftop but also space for tending vegetables.

FORM PROCESS

상처 입은 자연
오염된 물 , 분뇨
+
옥상 위 휴식공간
+
치료 하는데 필요한 에너지
(태양열 에너지)

총 네 가지 공간 필요

간단히 세워 놓을 수 있는
정팔면체

wounded nature
contaminated water,
manure
+
rooftop resting space
+
Energy required for
healing
(solar heat energy)

requires a total of 4
types of spaces

regular octahedron for
easy lift

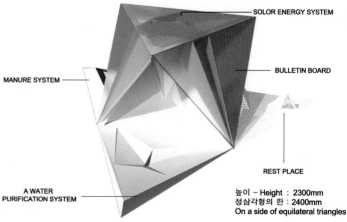

SOLOR ENERGY SYSTEM

BULLETIN BOARD

MANURE SYSTEM

REST PLACE

A WATER
PURIFICATION SYSTEM

높이 - Height : 2300mm
정삼각형의 한 : 2400mm
On a side of equilateral triangles

설치된 정팔면체는 구급상자 내에 필요한 기능들이 집약되었다 펼쳐지는 것과 같이, 자연을 치료할 수 있는 4 가지 시스템들이 펼쳐지는 구조로 계획되어 있다.

The installed regular octahedron is integrated with the functions required within the first aid kit. It is planned as a structure of unfolding the 4 system types that can heal nature.

PROGRAM PROCESS

01 SOLOR ENERGY SYSTEM

연료 감응형 태양전지
Fuel induced solar battery

생활 하수를 끌어올려 정수할 수 있는 에너지와 분뇨를 끌 어올릴 수 있는 에너지 생성

first aid kit가 스스로
작동할 수 있는 에너지

pulling sewage to form energy for pulling energy and waste for purification

energy for enabling independent operation of first-aid kit

02 REST PLACE

REST

ELECTRICITY

자연 친화적 형태로 잔디와 최소 한으로 닿게 디자인함. 기대어 앉아 풀위에 앉는 느낌을 받도록 했다.

designed in nature-friendly form to minimize touch with grass. designed to feel as if leaning and sitting on grass.

---

03 A WATER PURIFICATION SYSTEM

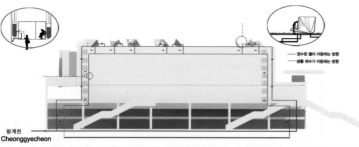

정수된 물이 이동하는 방향
생활 하수가 이동하는 방향

청계천
Cheonggyecheon

1층에서 4층까지의 상가와 5층부터 9층의 사무실에서 나온 생활하수들을 옥상으로 끌어올려 Frist Aid Kit로 정화 시킨다. 정수된 물은 옥상을 이용하는 사람들이 마실 수도 있으며, 남쪽에 위치한 청계천으로 흐르도록 할 수 있다.

pull sewage from commercial areas of 1st to 4th floors and office areas of 5th 9th floors to rooftop to purify through First Aid Kit. Purified water can be served to people using rooftop, or be flown to Cheongkye Creek located south.

자갈
gravel

모래
sand

활성탄
activated
carbon

여과지
filter paper

DRINKING   REUSING

AQUARIUM

양 쪽의 통로를 타고 온 생활하수는 가운데로 모 여 정수된 물이 어렵게 한 방울씩 나오게 됨으로 서 물의 소중함을 느낄 수 있다.

Sewage flown through routes from both sides gather to the center to create water with difficulty, providing realization regarding the value of water.

---

04 NANURE SYSTEM

가축분뇨, 인분 night soil	축분, 인분 저장탱크 tanks for storing night soil	가스 저장 탱크 tanks for storing gas	가스, 열, 전기 gas, heat, electricity
메탄가스 CH4	메탄가스 발효 Fermentation	가스배관 Gas piping	

냄새나는 인분과 가축의 분뇨에서 메탄가스를 제거하면 검은색 가루가 나오는데 냄새가 사라진 검은색 가루는 광장히 좋은 유기 질 퇴비가 되어 걸러진 메탄가스는 전기로 전환이 된다. 인분과 분뇨를 바이오 에너지로 활용한다면 화석연료의 사용이 줄어들 어 지구온난화를 완화할 수 있고 화학비료소비를 줄일 수 있다.

Removal of methane gas from smelly excrements and livestock manure forms black powder. This smell-removed black powder develops into extremely good organic compost while filtered methane gas is converted to electricity. Using excrement and manure as bio energy can reduce use of fossil fuels to improve global warming and decrease consumption of chemical fertilizer.

메탄가스 이동통로
분뇨에서 비료의 이동통로

NATURE   ELECTRICITY
VEGETABLE GARDEN

끌어올린 분뇨를 통에 담아 가스를 뽑아내고 남 은 비료를 모아놓고 사람들이 텃밭에 직접 뿌릴 수 있도록 하여 스스로 치료하고 식물을 돌볼 수 있도록 한다.

Pulled manure is placed into container to extract gas and collect remaining fertilizer to enable people to independently scatter on vegetable garden to facilitate independent healing and take care of plants.

FIRST – AID KIT에서 나온 물과 비료를 함께 사용하여 텃밭을 가꾸고 텃밭에서 나온 작물을 팔아 기부할 수 있도록 한다.
Water and fertilizer flown out from FIRST - AID KIT are used to tend vegetables and donations are made by selling crops grown from vegetable garden.

정삼각형으로 된 정 팔면체가 하나에서 여러 개로 합쳐졌을 때 기능은 증폭된다. 휴식공간을 여러 개 합쳐 여러 사람들과 편하게 쉴 수 있으며 누구나 조작을 손쉽게 할 수 있다.
Functions are amplified during the combination of several regular octahedrons, which are formed through regular triangles. The combined resting places enable various people to rest in comfort and can be easily operated by anyone.

SPACE SUBDIVISION

사람의 삶 자체가 물을 오염시킬 수 있는 소지를 가지고 있다. 가정에서 쓰고 버리는 물과 화장실에서 나오는 물, 공장과 사업장에서 버려지는 물, 소, 돼지 등의 가축을 기르는 데서 나오는 물, 논밭에서 농약과 비료가 섞여 나오는 물, 비가 내리면 도로에서 흘러내리는 물, 골프장에서 흘러나오는 농약 섞인 물, 야영지, 낚시터, 유원지에서 버려지는 음식 찌꺼기, 기름 찌꺼기 등등 일일이 열거하기 조차 어렵다. 그러나 물은 섞이고, 가라앉고, 퍼지는 등의 자정 능력(스스로 깨끗이 하는 힘)이 있어 어느 정도는 깨끗해진다. 그러나 이 자정 능력의 한계를 넘어설 때에 물은 심각하게 오염된다.

Human life itself possesses the aptitude for contaminating water. Water used and wasted from households, water from toilets, water wasted from factories and business sites, water from fostering livestock such as cows and pigs, water mixed with agricultural chemicals and fertilizer from farmland, water flown from roads during rain, water mixed with chemicals flown from golf courses, food and oil waste thrown away at camping grounds, fishing places, and resort areas, it is impossible to list them all. However, water also possesses the self-purifying power (ability to independently cleanse itself) of mixing, sinking, and spreading to become clean to a certain degree. However, water becomes seriously polluted when the limitations of such self-purification ability are exceeded.

우리나라 배출해역 폐기물 종류별 투입량
Waste input from discharged waters
in Korea according to type

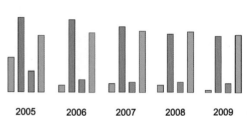

- 분뇨
- 가축 분뇨
- 폐수
- 음식물류 폐기물 폐수

2005 2006 2007 2008 2009

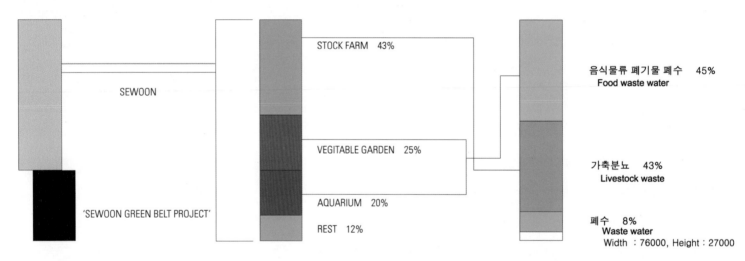

SEWOON

'SEWOON GREEN BELT PROJECT'

STOCK FARM   43%

VEGITABLE GARDEN   25%

AQUARIUM   20%

REST   12%

음식물류 폐기물 폐수   45%
Food waste water

가축분뇨   43%
Livestock waste

폐수   8%
Waste water
Width : 76000, Height : 27000

해양 투기 종합 관리 시스템에 의해 2009년 우리나라 배출해역 폐기물량에 대해서 알아보았다. 음식물류 폐기물 폐수가 44.9%로 가장 많은 량을 차지했으며 그 뒤를 가축 분뇨가 약 43%, 폐수가 약 8%, 분뇨가 약 5%를 차지했다. 이를 토대로 세운상가의 옥상위에 음식물 류 폐기물 폐수를 생활하수로 보고 45%를 수족관과 텃밭으로 사용하고 가축분뇨의 43%를 목장으로 사용하기로 했다. 폐수 8%와 분뇨 4%는 휴식공간이 될 예정이다.

Waste quantity of discharged waters in Korea of 2009 has been researched through the marine speculation synthesized control system. Food waste water occupied the largest quantity with 44.9%, livestock waste was measured as approximately 43%, waste water was measured as approximately 8%, and night soil was measured as approximately 5%.Based on such data, food waste water on rooftop of Sewoon Plaza will be regarded as sewage, and 45% will be used as aquarium and vegetable garden while 43% of livestock manure will be used as farm. 8% of waste water and 4% of manure will be used as resting place.

TRAIL

산책로는 높이를 다르게 하여 재미있게 표현한다.
Height of trail is differentiated to achieve playful expression.

STOCK FARM

높이를 600mm로 두어 가축들의 영역을 한정한다.
가축들은 정수된 물과 분뇨를 이용해서 경작한 작물들을 먹고 산다.
다 자란 가축들은 사회에 기부된다.
Height is adjusted to 600mm to restrict area of livestock. Livestock lives on cultivated crops by using purified water and manure. Full-grown livestock is donated to society.

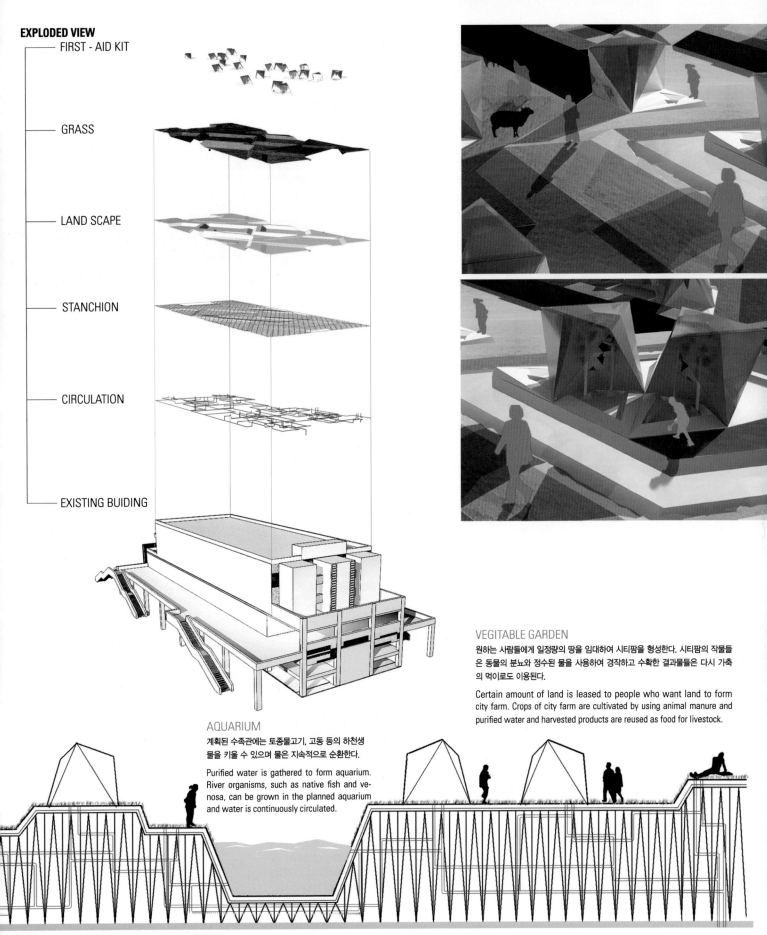

EXPLODED VIEW

- FIRST - AID KIT
- GRASS
- LAND SCAPE
- STANCHION
- CIRCULATION
- EXISTING BUIDING

VEGITABLE GARDEN

원하는 사람들에게 일정량의 땅을 임대하여 시티팜을 형성한다. 시티팜의 작물들은 동물의 분뇨와 정수된 물을 사용하여 경작하고 수확한 결과물들은 다시 가축의 먹이로도 이용된다.

Certain amount of land is leased to people who want land to form city farm. Crops of city farm are cultivated by using animal manure and purified water and harvested products are reused as food for livestock.

AQUARIUM

계획된 수족관에는 토종물고기, 고동 등의 하천생물을 키울 수 있으며 물은 지속적으로 순환한다.

Purified water is gathered to form aquarium. River organisms, such as native fish and venosa, can be grown in the planned aquarium and water is continuously circulated.

BACKGROUND

MBC 다큐 프로그램에 방송 출연 뒤 많은 관심을 받고 있는 승가원 장애아동시설. 관심과 지원이 끊이지 않는 승가원이지만 좁고 오래된 건물에서 생활하는 아이들에게 전혀 접해보지 못해오던 옥상 공간을 제안하여 진정한 아이들을 위한 공간을 만들고자 한다.

The Seungawon Facility for Disabled Children has been attracting much attention following its ap-

The Seungawon

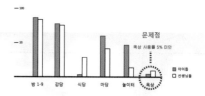

pearance on television via an MBC documentary. Though it has always been the subject of support and sponsorships, we plan to present a rooftop space previously unavailable to the children living within the old and cramped building, in order to provide them with a space truly of their own.

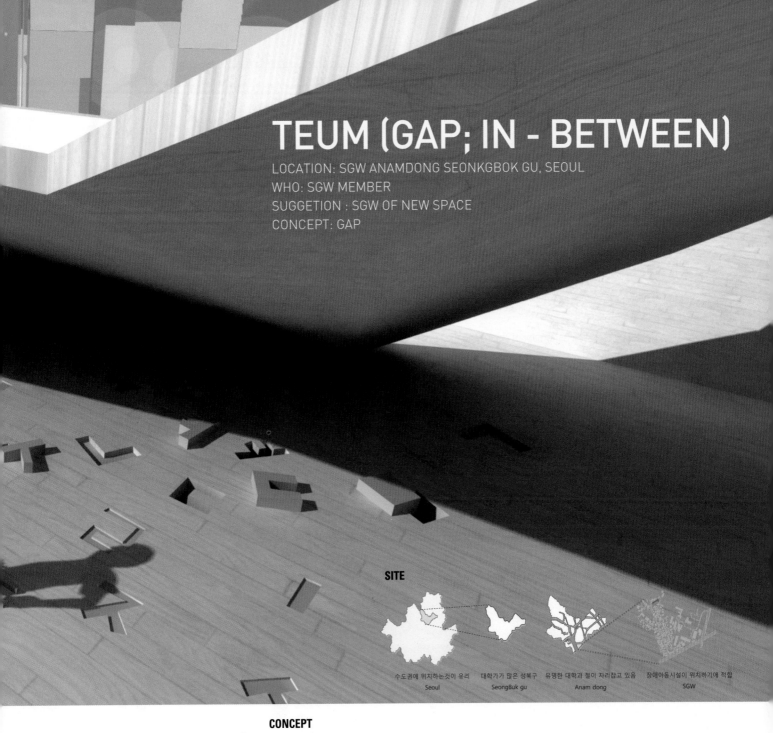

TEUM (GAP; IN - BETWEEN)

LOCATION: SGW ANAMDONG SEONKGBOK GU, SEOUL
WHO: SGW MEMBER
SUGGETION : SGW OF NEW SPACE
CONCEPT: GAP

SITE

수도권에 위치하는것이 유리   대학가가 많은 성북구   유명한 대학과 절이 자리잡고 있음   장애아동시설이 위치하기에 적합
Seoul   SeongBuk gu   Anam dong   SGW

CONCEPT

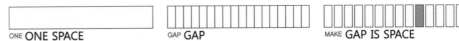

ONE **ONE SPACE**   GAP **GAP**   MAKE **GAP IS SPACE**

"틈" 이라는 사이사이의 새로운 공간을 제시해 주어 강당이라는 공통된 공간에서 생활하던 아이들은 자신의 성향에 맞는 틈의 공간을 가질 수 있게 된다. 그 공간 또한 틀에 박힌 공간에서 생활하던 아이들에게 닫힌 공간이 아닌 열린 공간이 됨으로써 승가원의 갑갑한 현실의 상황을 벗어 날 수 있는 탈출구가 된다.

By introducing new Teum(gap; in-between) spaces, the children who had lived in the commonly shared space of the auditorium, were now able to possess their own teum spaces fitting their individual characters. These spaces are also not the closed and fixed spaces that the children had previously lived in, but are open spaces that provide exits to the sometimes suffocating existence of living in Seungawon.

SGW SUGGESTION

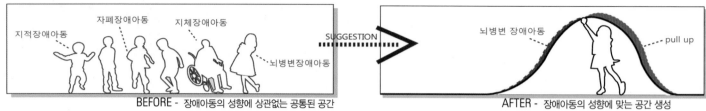

지적장애아동 자폐장애아동 지체장애아동 뇌병변장애아동

SUGGESTION >

뇌병변 장애아동 pull up

BEFORE - 장애아동의 성향에 상관없는 공통된 공간

AFTER - 장애아동의 성향에 맞는 공간 생성

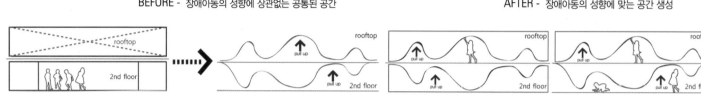

KEYWORD

RHYTHM 〰〰〰

틈의 반복과 반복을 통한 리듬의 형성 The Repetition of Teum, and the resulting generation of Rhythm

gap - repeat - make rhythm

gap - repeat - make rhythm
SGW - repeat - MAKE SPACE

틈의 반복을 통해 새로운 리듬의 형성이 가능해지며, 좁았던 공간에서의 틈 반복이 새로운 공간을 만들어낸다. 그 공간은 지루하기만 했던 승가원의 공간이 아닌 리듬을 가진, 리듬이 입혀진 틈 공간으로 형성된다.

The repetition of teum allows the generation of new rhythms and as the repeated teum formulates new spaces within the previously cramped spaces, these spaces are transformed from the dull spaces of Seungawon into teum spaces that possess a rhythm and are applied with rhythm.

DESIGN PROCESS

PART 1 . ELEVATION PROCESS
틈이 반복되는 리듬속에 필요한 공간의 도출
Deriving the sapce needed in the rhythm of repeate teum(GAP)

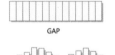

GAP

REPEAT

PULL UP ↑

PULL UP

MAKE SPACE

MAKE LINE

PART 2 . FLOOR PROCESS
평면에 얹어진 리듬으로 틈 공간의 시퀀스 발생
A sequence of teum space is generated from the rhythm imposed on the plane.

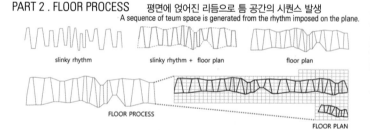

slinky rhythm

slinky rhythm + floor plan

floor plan

FLOOR PROCESS

FLOOR PLAN

PART 3 . RHYTHM PROCESS
필요에 따른 틈의 공간을 들어올리다.
The needed teum space elevated.

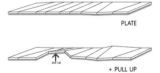

PLATE

PULL UP

+ PULL UP

NEW SPACE

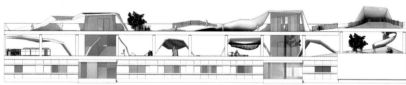

rooftop | funtion 1

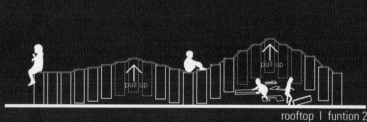

rooftop | funtion 2

LAYERS DIAGRAM

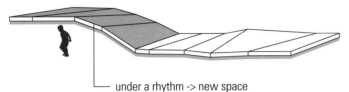

under a rhythm -> new space

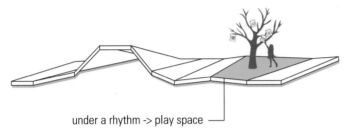

under a rhythm -> play space

리듬은 의미없는 리듬이 아닌 공간마다의 의미와 특징을 가지고 있다.

No rhythm is meaningless. All rhythm has its own meaning and characteristics according to the space it occupies.

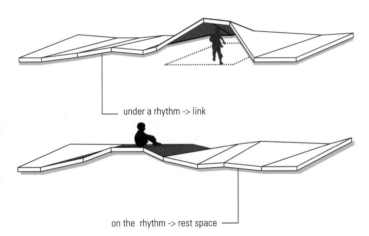

under a rhythm -> link

on the rhythm -> rest space

장애아동의 특성에 맞는 연결부분을 제공하여 불편요소를 줄인다.

Reducing factors for discomfort by providing links that agree with the physical properties of disabled children.

LAYER

리듬의 반복, 연속을 통한 공간의 생성.
The repelition of teum, and the resuling generation of Rhythm

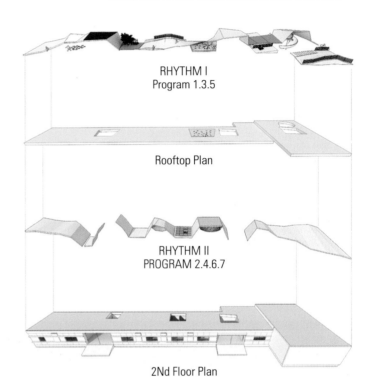

RHYTHM I
Program 1.3.5

Rooftop Plan

RHYTHM II
PROGRAM 2.4.6.7

2Nd Floor Plan

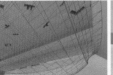

a rhythm -> new space

under a rhythm -> link

under a rhythm -> play space

on the rhythm -> rest space

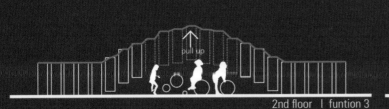

2nd floor | funtion 3

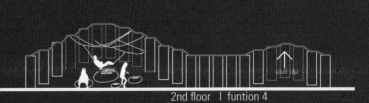

2nd floor | funtion 4

SPACE PROGRAM   지적장애아동을 위한 프로그램

· Making Shadow Pictures

지적장애아동의 적극적이고 효과적인 의사소통을 위한 놀이공간. 자신의 심리상태의 표정앞에서 그림자를 만들며 자신만의 표정을 찾는다.

A play space for the active and efficient communication of children with intellectual development disorders. They can produce shadows in front of what they find to be the face corresponding to their emotional state.

· Striking Energy

현재 승가원의 옥상은 solar cell이 반 이상을 차지하고 있어 아이들을 위한 옥상으로 활용되지 못한다. 아이들에게 알맞은 에너지로 변환하여 아이들이 자연스럽게 뛰고 즐기면서 진동에너지를 얻어 아이들을 치료한다.

Over half of the current rooftop of Seungawon is occupied by solar cells, which is why it had not been utilized as a space for the children. We plan to replace these cells with vibration energy harvesters, which will aid in treating the children as its installation will actually encourage their running about and enjoying themselves.

뇌병변장애아동을 위한 프로그램

· Put it Hangle

인지능력이 부족한 아이들에게 양손을 사용함과 동시에 직접 한글을 끼워맞추게 함으로써 지능의 발달과 재미적 요소를 제시해준다.

Inducing The Usage Of Both Hands In Children With Deficient Cognitive Abilities, And Having Them Fill In The Hangul Blanks Themselves, This Activity Provides The Opportunity For Fun And Development Of Intelligence At The Same Time.

· Folding Furniture

장애아동의 신체 특성과 자신이 원하는 가구를 만들 수 있는 곳으로 아이들의 창의력과 호기심을 자극하는 folding 가구 공간이다.

A space where disabled children can fold their own furniture to fit the needs dictated by their physical attributes, which stimulates their creativity and curiosity in the process.

making shadow pictures   striking energy   put it Hangle   folding furniture

Making shadow pictures   Stair   Striking energy

Folding furniture   SGW attic

238

SGW Attic

활동적인 아이들에게 호기심 가득한 공간을 제공하여, 자신만의 공간이 필요하거나 좀 더 새로운 틈의 공간을 원하게 될 경우 사용하여 아이들에게 자극을 주는 공간이다.

By providing a curious space for active children, we hope to stimulate children with a need for their own private space or some new teum spaces.

Sliding Puzzle

뇌병변장애아동의 양손을 사용할 수 있는 슬라이딩퍼즐을 바닥에 넣어 퍼즐을 밀어내고 물건을 찾아냄으로써 가장 효과적인 놀이치료를 기대한다.

Children with brain lesion disability use both of their hands to put sliding puzzles in the ground, push them about, and find items, which is expected to be a most effective form of play treatment.

Playing Mesh

하늘위에 누워있는 듯한 그물위에서 아이들은 자신의 신체에 맞는 공간을 만들어가고, 그 위에 한글이 끼워맞춰지면서 매번 새로운 하늘 공간이 된다.

Lying on a net that gives the sense of being in themddle of the sky, the children form spaces that fit theirown bodies, and as Hangul characters are put on top of the net, a new view of the sky is provided for every time the children use the net.

Photo Tree

사진찍기를 좋아하는 아이들이 자신이 찍은 사진을 공유하면서 의사소통을 자연스럽게 할 수 있으며, 나뭇잎을 채워가는 재미와 추억을 만들어 나가는 공간이 된다.

The children who enjoy taking pictures can communicate by sharing their photographs, and in the process of filling the leaves, this space can also be filled with memories and fun experiences.

틈과 틈,
사이사이의 공간마다 아이들의 성향에 맞는 공간을 제시한다.

Provide spaces that fit the children's characters in every in-between teum space.

놀이와 치료에서는 관계가 중요하며 최종적 목표는 정서적 성장이다.
In play and treatment the importance of relationship is paramount, and the ultimate goal of these activities is emotional growth.

즉, 나는 누구이고 나는 무엇이든지 할 수 있다는 확인감과 자아존중감을 고취시키는 것이 치료의 목표가 된다.
Thus, the aim of treatment is to inspire a sense of confidence and self-respect that allows the subject to know himself and believe that he can do anything.

승가원 장애 아동 생활 기준
지적장애아동, 뇌병변장애아동, 지체장애아동
시각, 청각, 언어, 자폐성 장애아동은 제외

구분		지적장애			뇌병변장애			자폐성장애		지체장애	
등급	합계	소계	1급	2급	소계	1급	2급	소계	1급	소계	1급
급수별 총원	72	44	40	4	26	26	-	1	1	1	1
남자	40	25	25	-	14	14	-	0	0	1	1
여자	32	19	15	4	12	12	-	1	1	0	0

뇌병변장애	뇌의 기질적 손상으로 인한 신체적, 정신적 장애로 독립적인 생활, 지적기능, 언어능력, 시각적 운동기능이 떨어짐.
지적장애	주변의 환경자극과 정보를 이용하는데 많은 어려움이 있고, 아동 스스로 환경에 적응하는 능력이 부족함.
정서장애	자신의 의지로는 통제하기 어려운 상태에 있어 공격적이며 파괴적인 행동이 아주 심함. 자책적인 감정이 지배적임.
청각장애	소리를 못듣거나 들어도 사람이 이해 할 수 없는 소리로 들려 이해하지 못함. 보통 수화를 사용.
지체장애	소아마비, 신체절단, 한센병 등으로 몸이 불편할 뿐. 생각하고 말하는 지적 능력은 정상인 장애인을 뜻한다.
시각장애	선천적 또는 후천적으로 시각능력이 없거나, 크게 떨어지는 장애인. 어떠한 경우가 되던 시각 장애는 일상 생활에 큰 불편을 겪는다.
자폐성장애	신체적, 사회적, 언어적으로 상호작용에서 이해 능력의 저하 일으키는 신경발달의 장애를 말한다.

지적장애아동을 위한 놀이치료 공간

지적장애아동, 선생님을 위한 연결공간 (stair)

뇌병변장애아동을 위한 놀이치료 공간

뇌병변장애아동을 위한 놀이치료 공간

지체장애아동을 위한 놀이치료 공간

지적장애아동을 위한 놀이치료 공간

틈 사이의 공간을 제시하여 공간부족의 문제점을 해결하고 항상 공통된 공간에서만 생활하던 아이들에게도 자신만의 공간을 가질 수 있음을 제안해 본다. 이 공간 안에서는 자신의 행동에 따라 공간이 가지는 의미가 달라질 수 도 있으며 2층과 옥상을 함께 느끼는 공간이 된다.

승가원의 선생님들의 공간 사무실

뇌병변장애아동을 위한 놀이치료 공간

지적장애아동을 위한 놀이치료 공간

지적장애아동을 위한 놀이치료 공간

지체장애아동을 위한 연결공간 (elevator)

지적장애아동을 위한 틈 공간 제안

주변의 환경자극과 정보를 이용하는데 많은 어려움이 있고, 아동 스스로 환경에 적응하는 능력이 부족함.

Intellectual disabilities: Experiences difficulty in utilizing environmental stimuli and information, and deters the ability to adapt to new environments without help.

intellectual disabilities

* 지적장애아동은 주변의 환경자극과 정보를 이용하는데 많은 어려움이 있고 아동 스스로 환경에 적응하는 능력을 기르기에는 아동이 갖고있는 전체 능력이 부족하다. 아동 개개인의 장애 정도와 특성에 따라 적절한 교육과 관련서비스를 제공하여 그들로 하여금 사회로의 통합과 생활의 자립을 이룰 수 있도록 해주어야 한다.

지적 장애아동에게 효과적인 치료 방법

치료 \ 효과 (%)	100	80	70	60
놀이치료	••••	•••	••	••
음악치료	•••	•••	••	•
미술치료	••••••	••••	•••	••
운동치료	•••	••	•	•

장애아동을 위한 미술치료

* 다양한 미술 활동을 통해서 아동의 지체된 발달 (인지적, 사회적 기능)을 촉진시킨다.

* 치료적 차원의 미술치료 : 장애특성으로 인한 다양한 심리적, 정서적 문제에 따른 다양한 적응 행동을 발생하게 될 때, 이러한 부적응 행동 소거를 목표로 미술치료를 진행 할 수 있다.

미술치료	••••••	••	••	••

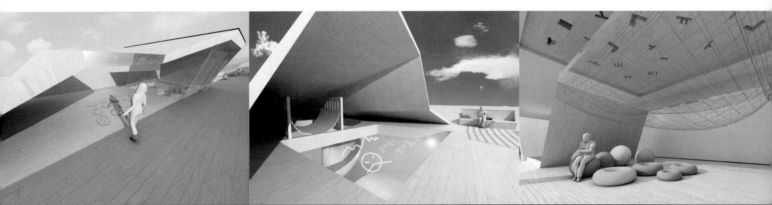

Making shadow pictures
자신의 심리상태의 표정앞에서 그림자를 만들며 자신만의 표정을 찾는다.

They can produce shadows in front of what they find to be the face corresponding to their emotional state.

Floor painting
종이나 칠판이 아닌 새로운 느낌의 바닥에 그림을 그림으로서 아이들의 집중력과 호기심을 유발하게 된다.

Painting on the floor of a new feeling of concentration and cusiosity of children will lead.

Play mesh
하늘위에 누워있는 듯한 그물위에서 아이들은 자신의 신체에 맞는 공간을 만들어간다.

Lying on a net that gives the sense of being in the middle of the sky, the children form spaces that fit their own bodies.

under a rhythm - new space

241

뇌병변장애아동을 위한 틈 공간 제안

뇌의 기질적 손상으로 인한 신체적,정신적 장애로 독립적인 생활,지적기능,언어능력,시각적 운동기능이 떨어짐.

This is a physical and mental disorder arising from organic damage of the brain, which impedes independent living abilities, intellectual functions, language abilities, and visual motor skills.

Brain lesion disorder

* 뇌병변장애아동은 왼쪽 또는 오른쪽의 팔, 다리의 마비인 경우로 주로 한쪽만 사용하기 때문에 성장에도 차이를 보여 팔, 다리가 짧거나 손과 발이 작은 경우가 많다. 감각이나 움직임에서 얻는 정보부족으로 인지능력 또한 부족하고 일반 아동에 비해 상대적으로 스트레스를 해소하지 못하며 오히려 내적으로 키우는 경우가 많다.

뇌병변장애 종류 분석

뇌병변 종류	뇌병변 장애 분석
편마비 뇌병변 장애	스트레스 해소 능력 부족, 성격이 급함, 정보부족으로 인지능력 부족, 양쪽 기능적 움직임이 어려움.
양마비(하지마비)의 뇌병변 장애	스트레스 해소 능력 부족, 뇌의 인지영역 손상 또는 정상적인 감각이나 움직임에서 얻는 정보 부족으로 인지 학습 필요.
사지마비의 뇌병변 장애	스트레스 해소 능력 부족, 기능적 움직임 뿐 아니라 목을 똑바로 가눌 수 없어 섭식, 배변, 눈맞춤, 언어기능, 균형감각, 인지능력 부족함.

뇌병변장애아동에게 효과적인 치료 방법

	미술치료	언어치료	운동치료	인지학습	스트레스 해소
편마비 뇌병변 장애					
양마비(하지마비)의 뇌병변 장애					
사지마비의 뇌병변 장애					

PUT IT HANGLE

인지능력이 부족한 아이들에게 양손을 사용함과 동시에 직접 한글을 끼워맞추게 함으로써 지능의 발달과 재미적 요소를 제시해준다.

Inducing the usage of both hands in children with deficient cognitive abilities, and having them fill in the Hangul blanks themselves, this activity provides the opportunity for fun and development of intelligence at the same time.

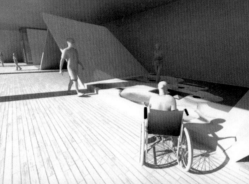

SLIDING PUZZLE

뇌병변장애아동의 양손을 사용할 수 있는 슬라이딩퍼즐을 바닥에 넣어 퍼즐을 밀어내고 물건을 찾아냄으로써 가장 효과적인 놀이치료를 기대한다.

Children with brain lesion disability use both of their hands to put sliding puzzles in the ground, push them about, and find items, which is expected to be a most effective form of play treatment.

SLIDE

신체가 자유로운 아이들이 2층과 옥상을 자유롭게 드나들 수 있으며, 이동이 불편하였던 아이들에게 재미있는 연결 통로가 되게한다.

Children with free use of their body can move between the second floor and the roof floor, while those whose movement was restricted are provided with a fun linking route.

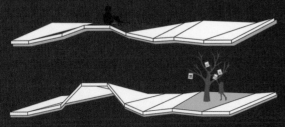

on the rhythm - rest space

on the rhythm - play space

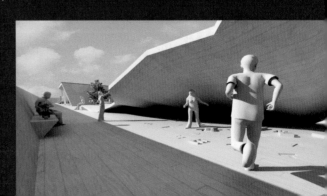

지체장애아동을 위한 틈 공간 제안

소아마비,신체절단,한센병 등으로 몸이 불편할 뿐 생각하고 말하는 지적 능력은 정상인 장애인을 뜻한다.

Includes polio, amputees, Hansen's etc. Only the body is affected with the intellectual facilities remaining normal.

Physical disabilities

* 승가원 장애아동 시설에서 지체장애를 가진 아동은 한 명 뿐이다. 바로 2000년 승가원에 입소한 11살 태호인데 태호는 태어날 때부터 양 팔과 허벅지부분 없이 태어났다. 심장과 폐가 약한 태호의 심폐기능을 강화하고 발로 생활하는 태호에게 몸통의 균형을 잡아주는 치료가 필요하다.

WEEKLY PROGRAM

장애아동의 성향에 맞는 공간을 형성해주어 자신의 특성에 맞는 공간 활용도가 나타난다.

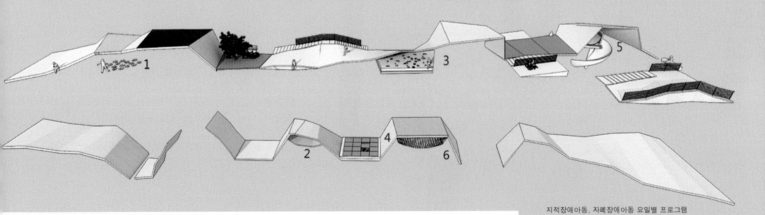

TRAFFIC LINE

승가원 장애 아동의 성향에 따라 공간을 이동하므로 장애 아동 성향에 맞는 동선이 형성

LINE 1 - SGW CHILD . 1   지적장애아동. 자폐장애아동
LINE 1 - SGW CHILD . 2   뇌병변장애아동. 지체장애아동

LAYER 1

LINE 1 - SGW CHILD . 1   지적장애아동. 자폐장애아동
LINE 1 - SGW CHILD . 2   뇌병변장애아동. 지체장애아동

LAYER 2

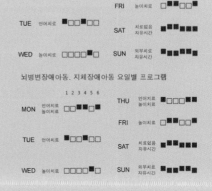

지적장애아동. 자폐장애아동 요일별 프로그램

뇌병변장애아동. 지체장애아동 요일별 프로그램

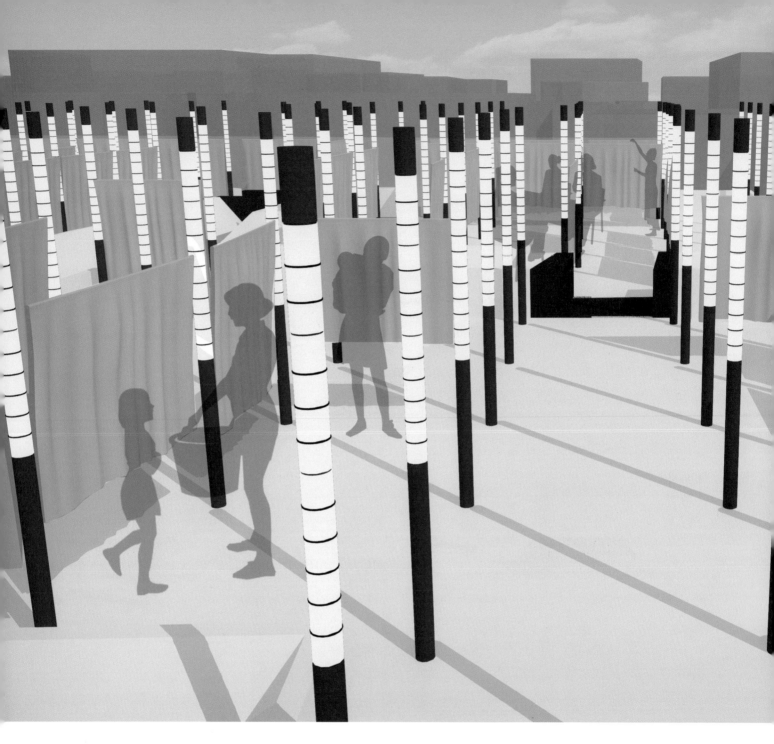

Present rooftop [현재 옥상의 모습들]

*Proposal of the rooftop space that will give
energy to dull routine of housewives.*

Patch a Quilt
주부의 지루한 일상에 활기를 안겨 줄 옥상 공간 제안

가정에 대한 헌신적인 자세는 강요되면서도 스스로를 돌보고 가꾸는 것에 있어서는 소극적이 될 수뿐이 없는 이 시대 한국의 어머니는 어쩌면 사회적 약자임에 틀림없다. 사회의 중요한 구성원임에 틀림없지만 가정이란 보이지 않는 경계에 갇혀 주위로부터 소외된 그들에게 지금 가장 필요한 것은 소통이다. 본 프로젝트는 옥상이란 감춰진 공간위에 가족과 이웃이 소통할 수 있는 장을 계획한다. 밀집된 집합주거의 거주환경에서 옥상은 적극적인 생활영역에 포함되지 못하고 부가적인 기만을 수용하는 소극적 존재에 지나지 않는다. 이러한 옥상 위에 최소한의 구조물을 설치함으로서 주부들의 화합과 소통을 이끌어 내고자 하는 것이 본 프로젝트의 목적이다.설치물을 사용하면서 자발적이고 자기조직적인 커뮤니케이션과 활동이 이루어지도록 유도한다. 그럼으로써 가정의 조력자로서의 주부가 아닌 활동의 주체로서의 주부들을 발견하고 생활에 활력을 얻을 수 있는 공간이 되기를 바란다.

The mothers of Korea in this age are forced to have devotion for the home, and be passive when it comes to looking after themselves. They are perhaps the socially disadvantaged. They are one of the important members of the society, but they are also alienated from the surroundings by being locked in invisible boundaries. For them, communication is what they need the most. This project plans the field where families and neighbors can communicate with each other, on top of the space hidden on the rooftop. In crowded housing living environment, rooftop is excluded from active living area and is nothing more than a passive space that accommodates additional deception. The purpose of this project is to install the minimum structures on the rooftop and induce harmony and communication among housewives. By using various structures, voluntary and self-organizing communication and activity are induced. By using such methods, this project hopes that housewives discover themselves as the principal agent of activity, instead of housewives, and obtain a tonic of life.

TARGET

봉건적 사회 분위기라고는 어디에서도 찾아 볼 수 없는 2010년 한국이지만, 대다수의 40–50대 주부들은 자신만의 시간과 공간을 갖지 못한 채 오직 가족과 자식들을 위해 가정에 봉사해야 한다. 이러한 주부들을 위해서 주부들만의 휴식공간과 다른 주부들과의 친목을 통해 그들 스스로 프로그램을 계획하고 그 안에서 영위할 수 있는 옥상공간을 제안하고자 한다.

Feudal social atmosphere is nowhere to befound in 2010's Korea, but majority of housewives in their 40's and 50's have to provide domestic services for families and children only, without having their own time and space. The rooftop space, where such housewives can plan their own program and manage internally through friendship with other housewives in their resting space, is proposed.

SITE

−면적: 약 20.5km2
−주소: 부천시 원미구 원미동
−거주인원: 약200명
　　　(건물당 4–5가구, 1가구당 3–4명)
−단지 내 옥상높이가 비슷한 밀집된 건물 10채.

-Area: approximately 20.5km2
-Address:Bucheon-city,Wonmi-gu,Wonmi-dong
-Population: approximately 200 (4-5 hous holds per building, 3-4 people per household)
-10 buildings within the complex with similar rooftop height

CONCEPT

Pixel 픽셀

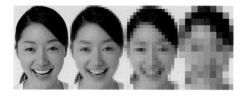

픽셀은 이미지를 구성하는 작은 점으로, 유사한 색을 가진 단일 픽셀들이 붙여 나가지면서 전체 모습이 경계지지 않고 연속된 화상을 형성하도록 하는 화상의 최소단위라는 개념을 가지고 있다. 이웃과의 경계를 완화시키고 연속된 주부공동체를 형성하도록 한다. 옥상에 픽셀을 삽입하는 것은 일차적으로 단순히 단지를 묶어내는 것이 아닌, 공간에 주부의 특성에 알맞은 프로그램을 삽입하여 픽셀처럼 맞붙어있는 공간이 물리적으로 연결될 뿐 아니라 옥상을 이웃 간의 정신적인 연계가 될 수 있도록 한다.

Pixel is a small dot that forms images, and has a concept of the image's smallest unit which forms the overall continuous, non-bordered images by attaching unit pixels. The boundary with neighbors are loosened, and a continuous housewives community is formed. Inserting pixels on rooftop is not limited to simply binding complexes. By inserting appropriate program for housewife's characteristic to the space, spaces that are next to each other like pixels are physically connected, and the rooftop becomes a mental connection among the neighbors.

Empty space

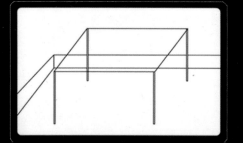

One person,

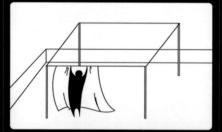

Drying........?

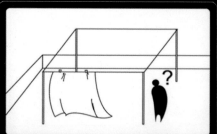

PROCESS

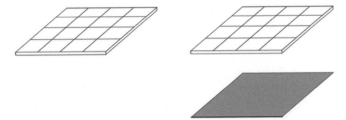 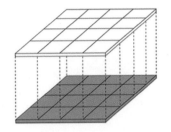 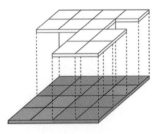

Step1

[옥상공간의 연결을 통해 빨랫줄에 의해 구획된 단지의 유기적 연결 및 통합]

Organic connection and integration of the complex divided by clothesline, through the rooftop space connection.

Step2

[입체계획을 통해 2차원적으로 구획된 옥상을 3차원적으로 계획함으로써 옥상과 주민, 옥상과 사회, 주민과 사회의 관계 맺기]

Forming relationships between the rooftop and residents, rooftop and society, and residents and society by 3-dimensionally planning the rooftop that is 2-dimensionally divided, through dimensional planning.

Step3

[개인중심의 공간구조였던 단일기능 옥상에서 주민과 공동사회중심의 다기능 복합 옥상 제안]

Proposing multi-functional rooftop that is centered on residents and communal society, different from existing single-function roof that was individual-oriented.

빨랫줄로 구획된 단지.	옥상공간의 통합.	옥상공간의 활용.	옥상공간을 통한지역주민의 연결.
The complex divided by clothesline	Integration of the rooftop space	Utilizing the rooftop space	Connecting local residents through the rooftop space

CREATE SYSTEM

일정한 격자패턴으로 장대(post)를 설치하고 장대에 빨랫줄을 묶는다. 빨랫줄에 각자의 집에서 가지고 온 이불이나 천 등을 널어 모두 함께 사용할 수 있는 공간을 만들어간다. 단 한 사람만으로는 공간을 만들기 어렵다. 하지만 사람들이 모이고 모여서 공간을 만들어 넓혀 간다. 공간을 만들 때 여러 사람이 각자의 이불을 가져와 널면 여러 개의 무늬가 모여 하나의 거대한 퀼트가 만들어질 수 있다. 이렇게 자발적으로 공간을 만든 주부들은 자연스럽게 자발적인 단체 활동을 하게 되며, 자기조직적 성격을 띈다.

Post is installed in a regular grid pattern and clotheslines are tied on the post. Blankets and clothes from each household are hung on the clothesline, creating a space that can be used by everyone. It is difficult to create such a space with only a single person. However, people gather and expand such a space. When creating the space, each person brings their own blanket and hangs them on the clothesline, and many patterns are collected, creating a massive quilt as a result. Housewives who have voluntarily created such a space is naturally participating in a group activity, and taking on self-organizing characteristic.

Some person,

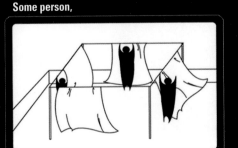

Be creative of space!

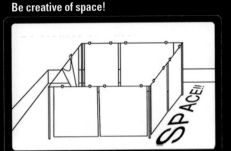

Self-organized activity.

Personal rooftop

Integrated rooftop

STRATEGY

빨랫줄과 천으로 만들어지는 공간과 그 안에서의 단체 활동을 통해 단지 내 주부 화합을 위한 옥상공간을 제공한다. 이는 사회에서 느낄 수 있는 우울감이나 소외감등을 해결해 준다.

Through the space that consists of clotheslines, clothes and group activities within it, the rooftop space for the harmony of housewives of the complex is provided. Such space solves the issue of depression and a sense of alienation that arise in society.

Flower bed

Warehouse

Water tank

Pole

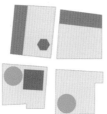

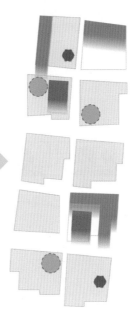

현재 사용 중인 옥상 상황
State of the rooftop currently in use.

공동공간형성으로 또 다른 공동 공간 만들기
Creating another joint space with the formation of joint space.

옥상에 분포되어 있는 사용하고 있었던 화단과 창고를 북쪽으로 밀집시키고, 사용하지 않는 물탱크와 기둥은 철거한다. 합쳐진 화단과 창고는 북쪽의 2채에서 공동으로 사용하고 확보된 8채의 옥상공간에 빨랫대를 설치해 공동공간으로 사용할 수 있도록 한다.

Move flower beds and warehouses that were being used on the rooftop to the north, and remove the water tank and poles. Combined flower beds and warehouses are jointly used in two northern buildings, and clotheslines are installed on the rooftop space of eight remaining buildings and used as a joint space.

Human scale

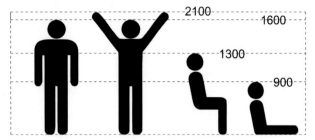

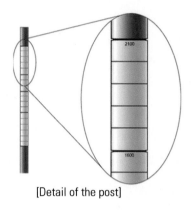

[Detail of the post]

[평균 신장 160cm 여성 기준] 빨랫대는 human scale을 근거로 나온 주요 치수에 줄을 맬 수 있게 했다. 공간의 쓰임에 따른 다양한 평면 변화. 화단과 창고를 합한 계단형 좌석. 계단에 앉아 영화보기. 다양한 공간을 만들 수 있는 격자 빨랫줄. 조명으로 활용 가능한 빨랫대.

[Based on female average height of 160cm] Tying strings was made possible on the laundry pole based on human scale measurements. Various plane changes according to space and usage. Step seat that combines flower beds and warehouse. Watching movie by sitting on the stairway. Grid clothesline that can generate various spaces. Laundry pole that can be utilized as the lighting.

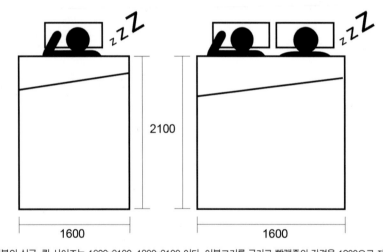

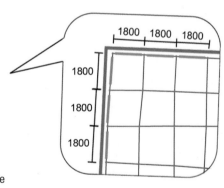

[Clothesline intervals]

이불의 싱글, 퀸 사이즈는 1600x2100, 1800x2100 이다. 이불크기를 근거로 빨랫줄의 간격을 1800으로 제안한다.

Blanket's single and queen sizes are 1600x2100 and 1800x2100. Based on the blanket sizes, clothesline intervals of 1800 is propose

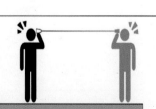

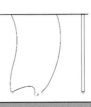

251

[Using a variety of spaces]

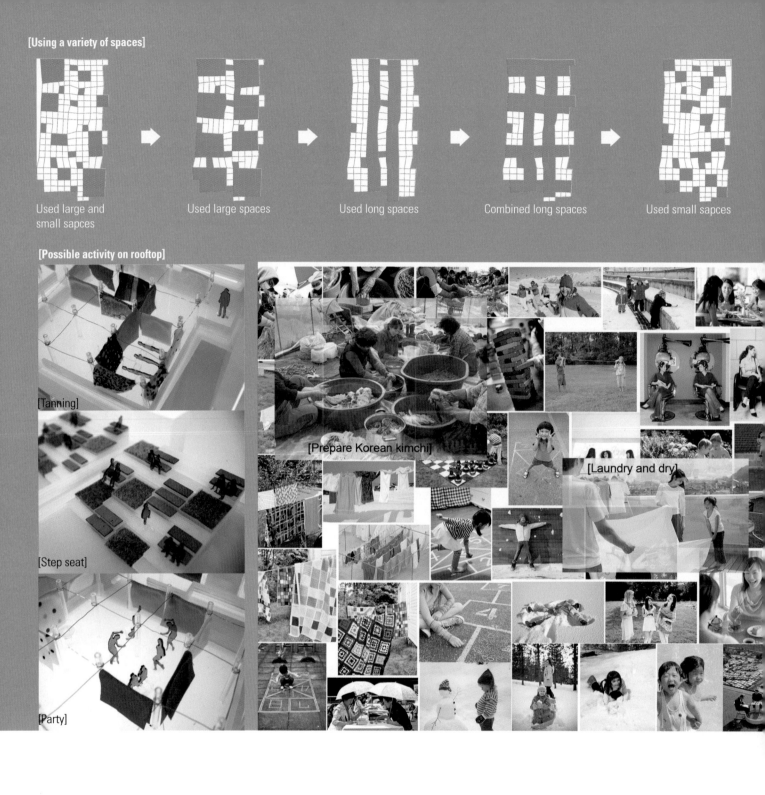

Used large and small sapces → Used large spaces → Used long spaces → Combined long spaces → Used small sapces

[Possible activity on rooftop]

[Tanning]

[Step seat]

[Party]

[Prepare Korean kimchi]

[Laundry and dry]

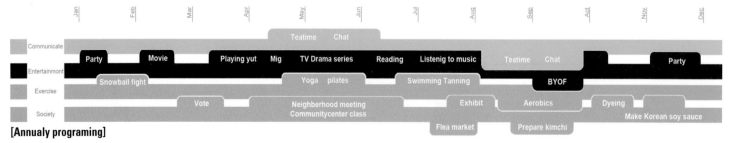

[Annualy programing]

	Jan	Feb	Mar	Apr	May	Jun	Jul	Aug	Sep	Oct	Nov	Dec
Communicate					Teatime Chat							
Entertainment	Party	Movie	Playing yut	Mig	TV Drama series	Reading	Listenig to music	Teatime Chat				Party
Exercise	Snowball fight				Yoga pilates		Swimming Tanning		BYOF			
Society		Vote		Neighborhood meeting Communitycenter class		Exhibit	Aerobics		Dyeing	Make Korean soy sauce		
					Flea market		Prepare kimchi					

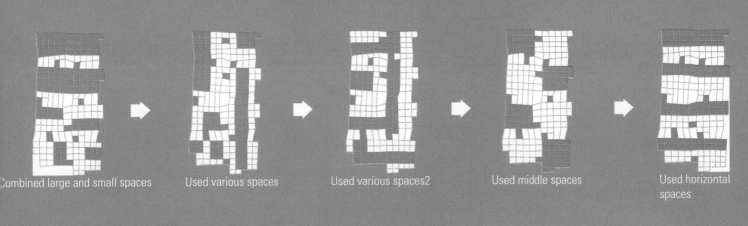

Combined large and small spaces → Used various spaces → Used various spaces2 → Used middle spaces → Used horizontal spaces

[Flower beds]

[Party]

[Family dinner]

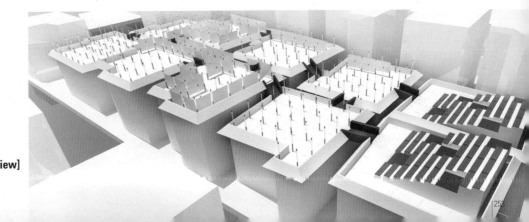

[Bird's-eye view]

길쌈놀이(Gilsamnolyi)

아티스트의 작업공간을 옥상에 짜다.

SITE ANALYSIS

BACKGROUND

요즘, 아티스트 레지던스 공간이 늘어나고 있는 추세이다. 하지만 그것은 좋은말로 아티스트 레지던스 공간이지 실제로는 아티스트들이 작업할 공간이 없어 생겨난 공간이다. 또한 철공소 사람들은 시대가 변하면서 대량생산이 생겨남으로 인해 생계에 위협을 느끼고 있다.

이러한 상황으로 봤을 때 아티스트는 공간적 약자로, 철공소 사람들은 사회적 약자로 보고 이들의 공존을 통해 사회적, 공간적 약자를 벗어날 수 있는 공간을 제시하고자 한다.

Nowadays, artist residence spaces are on the rise. But it is the artist residence space by good words but was created because there was no space for artists actually. Also, ironworkers' livelihoods are threatened because of the mass production, as times have changed.
In this situation, by considering that artists are the spatial weak and ironworkers are the social weak, the space which can get out of the social and spatial weak intends to be suggested through the coexistence of them.

SITE ANALYSIS

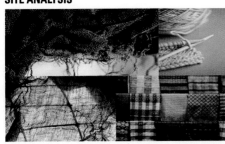

천은 씨실과 날실이 엮여 하나의 형태가 만들어진다. 그러므로 우리는 기존에 살고있는 철공소 사람들을 씨실로, 아티스트들을 날실로 보고 그 둘을 엮어 옥상에 하나의 천을 만들고자 한다.

The fabric makes one form by weaving fillings and warp threads. Therefore, we intend to make one fabric on the roof by considering that ironworkers living originally are fillings and artists are warp threads and weaving them.

KEYWORD

씨실, 날실 > 중첩. 그리드 Knitting-connection, repeat	절단 Cutting	팽창,수축 – 구멍,주름 Knitting-connection, repeat	확대 Enlargement	뜨개질 – 연결,반복 Knitting-connection, repeat

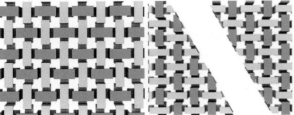  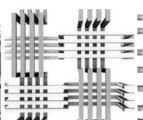 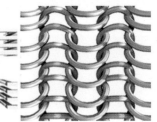

녹지공간 (green space)
경사 (slope)

중간중간 녹지공간을 틈틈히 제공함으로 도시의 삭막함을 벗어 날수 있게 하며,
지붕 위의 경사 그대로를 사용함으로 각각의 지붕의 느낌이 전해져 하나의 재미를 부여 한다.

TEXTILES VARIETY

Textile　Flexure　Wrinkles　Geometry　Drape　Expansion　Transformation　Shrinkage　Neckline　Separation　Expansion1　Expansion2　Expansion3　Expansion4

공간이 한정되어 있는 실내를 벗어나 실외에서 옥상이라는 특성을 갖고 자유롭게 작업을 한다.

위 아래가 커뮤니티라는 실들로 연결되어 서로간의 구분을 없애고 함께 공생하며 더없이 좋은 사회가 된다.

SPACE PROGRAMING

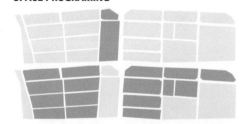  

우리는 블록을 하나의 개체로 보고, 하나의 개체가 모여 공간을 형성했다. 그리고 그 공간을 모아 핵심적인 공간을 도출 했으며, 그 핵심적인 공간 안에 커뮤니티 및 전시 공간을 형성했다. 그렇게 블록단위로 모여진 옥상 공간을 아티스트가 작업만 할 수 있는 작업실 및 휴식 공간을 제공하였고, 블록들 중 중심이 되는 공간에 전시 공간 및 커뮤니티 공간을 형성했다. 씨실은 옥상에 가로로 놓여진다.

그렇게 놓여진 씨실을 연결해 블록과 블록의 브릿지를 만들었다. 그렇게 만든 브릿지를 통해 블록의 가로라인끼리 하나의 그룹이 형성된다. 이렇게 만들어진 그룹에 하나씩 계단과 안내데스크를 만들어주었으며 일반 사람들이 전시를 구경할 수 있도록 전시공간을 위한 계단을 만들었다. 전시공간의 동선은 씨실이 팽창되면서 세 개의 공간이 형성되고, 그 형성된 공간 안에 또 팽창이 일어나 동선을 만들어준다.

전시공간의 동선은 주로 S자식이며, 그 씨실들이 안으로 들어가면서 바깥쪽에 휴식공간을 제공 해준다. 그리고 블록의 동선은 크게 실내 작업 공간, 외부 작업 공간, 휴식공간으로 나눠진다. 실내 작업 공간 앞에 외부 작업 공간이 있고, 그 공간과 공간 사이에 길이 만들어지면서 휴식공간이 생겨난다.

We formed the space with one individual, by considering the block is one individual. And core space was drawn by collecting the space and community and exhibition space were formed in the core space. With the rooftop space collected under the unit of block like that, workroom and resting space that artists can only work were provided. And among the blocks, exhibition and community spaces were formed in the central space. Fillings are laid on the rooftop crossly.

By connecting those fillings, the bridge among the blocks was made. Through the bridge, one group is formed by width lines of block. One stair and one information desk were made for each group and stairs were made for exhibition space so that general people can see the exhibition. In the m ovement of exhibition space, three spaces were formed by expanding fillings. And in the space, the movement was made with expansion again.

The movement of exhibition space is mainly the type of 'S' letter and resting space is provided outside by inserting those fillings to inside.

And the movement of block is divided into indoor working space, outdoor working space and resting space largely. There is outdoor working space in front of indoor working space and the resting space is created by making the street between the spaces.

GALLERY

PLAN

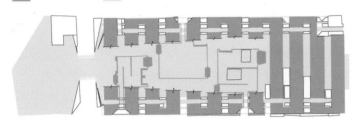

■ 씨실 filling　□ 날실 warp threads

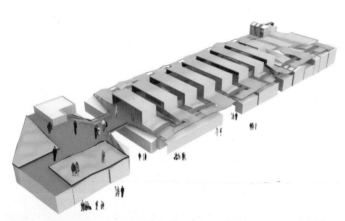

EXPLODED VIEW

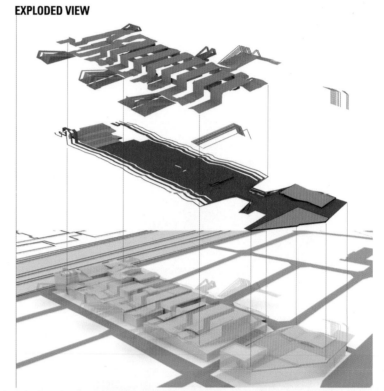

GALLERY DIAGRAM

  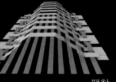  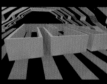

씨실,날실　　　　분리　　　　팽창　　　　공간생성　　　　동선생성

GALLERY BRIDGE DIAGRAM

 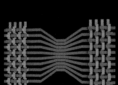 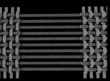

씨실,날실　　　　주름생성　　　　수축　　　　연결　　　　날실 팽창

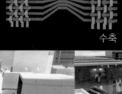   

BLOCK

PLAN

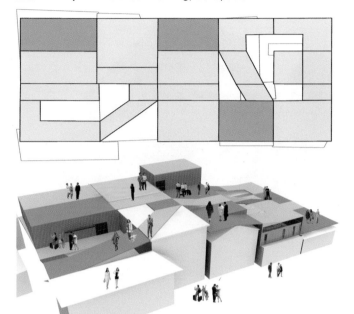

EXPLODED VIEW

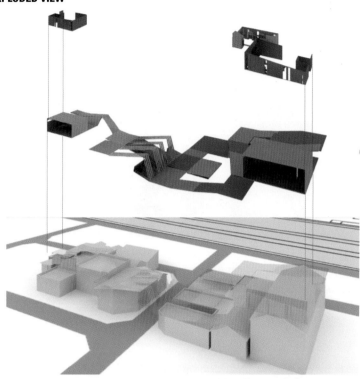

BLOCK DIAGRAM

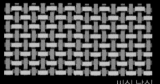
씨실.날실

절단

높낮이

길생성

잔디

BLOCK BRIDGE DIAGRM

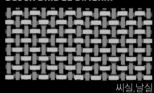
씨실.날실

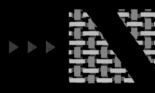
절단

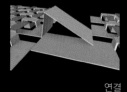
연결

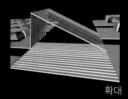
확대

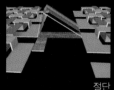
절단

UNIT-S
개인이 작업하는 스튜디오

PLAN

TOILET

WORK SPACE

REST SPACE

이 공간은 개인이 작업하는 공간이므로 팽창을 이용해 책상을 만듦.
This space is for individual working, so the desk is made
by expansion.

유닛의 한 부분을 뚫어 주고 전창을 줌으로 채광을 받되, 프라이버
시 보호가 될 수 있도록 하며, 그 외 부분 창은 작업을 하는 도중 생
겨나는 공기오염을 환기 시켜주고 바깥 공기와 안 공기가 잘 순환
되게 하여 작업함에 있어 좋은 환경을 부여해주기 위함이다. 개인
이 작업하는 이 공간은 금속을 재료로 하고 작업함에 있어 주위의
부수적인 것들에 방해가 되지 않게 공간을 분할하며, 원활한 작업
이 이루어질 수 있도록 넉넉한 수납 공간과 휴식을 취할 수 있는 휴
식 공간을 배치 하였다.

DESIGN PROCESS

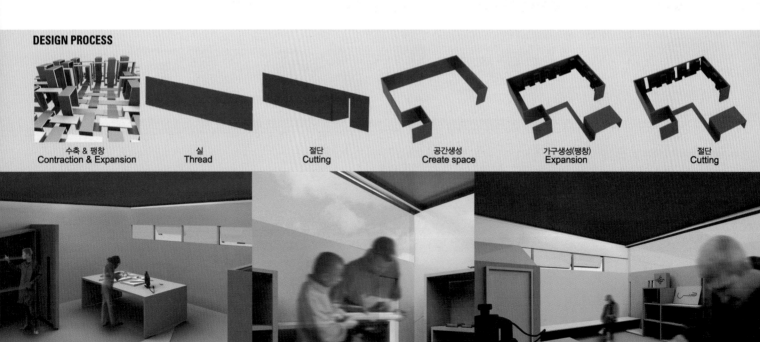

수축 & 팽창
Contraction & Expansion

실
Thread

절단
Cutting

공간생성
Create space

가구생성(팽창)
Expansion

절단
Cutting

UNIT-M 소그룹이 작업하는 스튜디오

PLAN

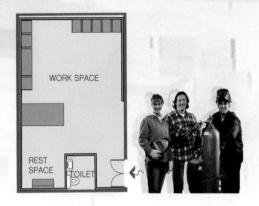

이 공간은 소그룹이 작업하는 공간으로 절단을 이용해 수납 공간을 늘렸음.

This space is for small group's working, so the storage space is increased by cutting.

소그룹이 작업하는 이 공간은 작업이 그룹으로 이루어 지기 때문에 서로 소통하며 작업에 임할 수 있도록 하며, 책상을 사용하되 바닥에서의 작업도 원활하게 할 수 있도록 넓은 공간을 확보 주었다. 이 공간 또한 작업 공간 외에 한쪽에 휴식을 취할 수 있는 공간이 배치 하였다. 유닛의 벽 높이를 장소에 따른 높이를 조절하여 공간을 더욱 적합하게 하며, 일률적인 높이의 지루함을 줄일 수 있도록 한다.

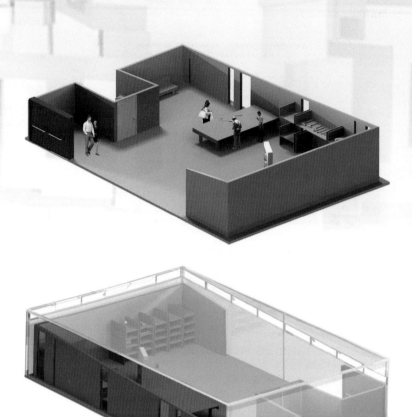

DESIGN PROCESS

| 절단
Cutting | 실
Thread | 절단
Cutting | 공간생성
Create space | 가구생성(절단)
Cutting | 절단
Cutting |

265

UNIT-L 대그룹이 작업하는 스튜디오

PLAN

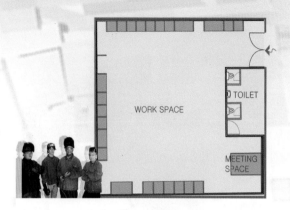

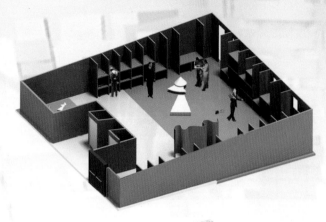

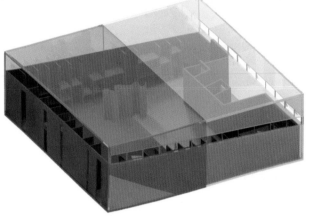

이 공간은 개인과 소그룹이 모여 공동작업을 하는 공간으로 팽창을 이용해 회의공간을 만듦.

This space is for the teamwork of individual and small group, so the conference space is made by expansion.

대그룹이 작업하는 이 공간은 많은 사람들이 공동으로 대형 조형물을 제작하기 때문에 책상보다는 넓은 공간을 확보하고 사방으로 수납 공간을 배치하여 손쉽게 공구를 가져 쓸 수 있도록 하였다. 다수의 사람들이 공동으로 작업하기때문에 간단한 회의를 할 수 있는 회의실을 배치하였다. 대 그룹 공간은 많은 사람이 편리하게 작업할 수 있도록 두 공간을 합쳐 하나의 공간으로 만들고, 유닛벽의 높낮이를 따라 부분 창문을 사선으로 구성한다.

DESIGN PROCESS

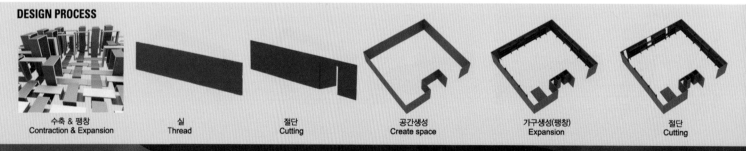

| 수축 & 팽창 Contraction & Expansion | 실 Thread | 절단 Cutting | 공간생성 Create space | 가구생성(팽창) Expansion | 절단 Cutting |

UNIT-I INFORMATION

PLAN

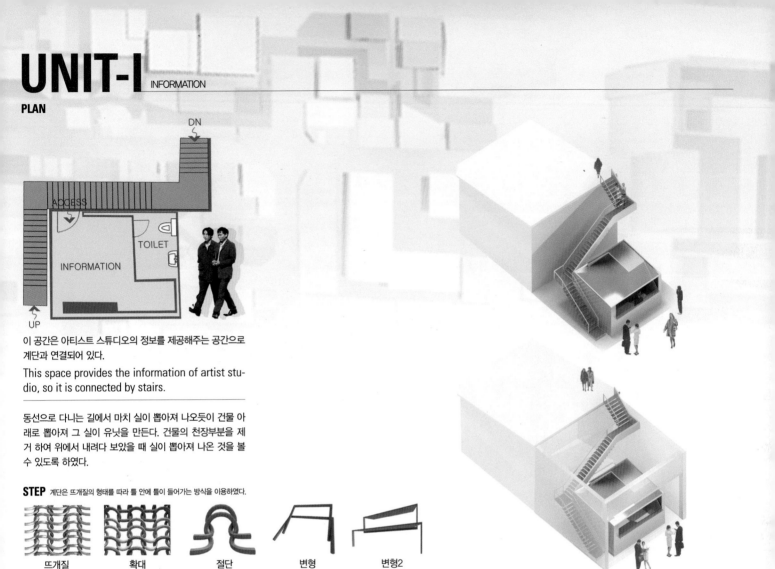

이 공간은 아티스트 스튜디오의 정보를 제공해주는 공간으로 계단과 연결되어 있다.

This space provides the information of artist studio, so it is connected by stairs.

동선으로 다니는 길에서 마치 실이 뽑아져 나오듯이 건물 아래로 뽑아져 그 실이 유닛을 만든다. 건물의 천장부분을 제거 하여 위에서 내려다 보았을 때 실이 뽑아져 나온 것을 볼 수 있도록 하였다.

STEP 계단은 뜨개질의 형태를 따라 틀 안에 틀이 들어가는 방식을 이용하였다.

뜨개질	확대	절단	변형	변형2

DESIGN PROCESS

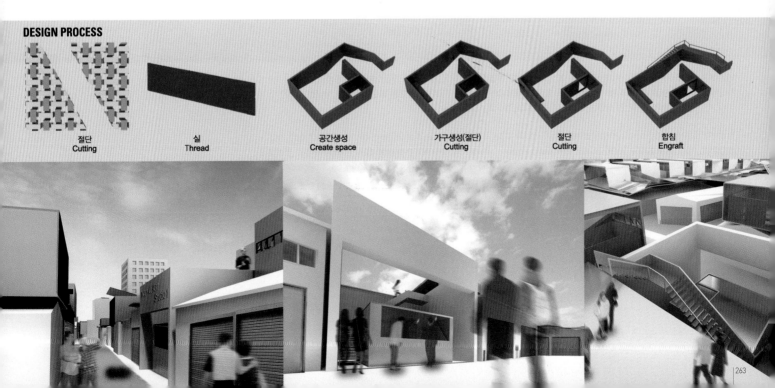

절단 Cutting	실 Thread	공간생성 Create space	가구생성(절단) Cutting	절단 Cutting	합침 Engraft

예전부터 우리나라는 '개인' 보다는 '우리' 를 강조 하였다. 우리집, 우리마을, 우리동네, 우리아빠, 우리동생....... '우리' 라는 단어는 어색하지않다. '우리 네' 는 예전부터 집 앞에 마당이라는 공간을 두었다. 마당이란 딱히 쓰임새가 정해진 공간은 아니다. 아이들의 놀이터가 되는가 하면 동네의 잔치를 치르는 장소로 사용되었다.

From old days, Korea has emphasized "we" more than "I". Our house, our village, our town, our dad, our brother, etc. The word "we" is not awkward at all. "Ours" has placed a space called Madang in front of the house from old days. Madang is the space with no specific purpose. Sometimes it becomes a playground for children, or a place of a party of the village.

Story about our selves

BACKGROUND

현대사회의 빠른 발전중에 있어 낡은
건물들은 어느새 우리에게 있어 그저
시대에 뒤 떨어져 없어져야 할 것 이
라는 생각을 가지게 되었다.

잘못된 시선으로 인해 시행된 재개
발로 삶의 자리를 잃어버리게되고
밖으로 나가거나 다시 들어올 수 없
는 사람들

주변과의 조화를 위해 재개발을 선
택 하였지만 그 곳의 살던 사람 들
의 문제 는 하나도 해결 되지 않는 문
제점만을 낳게 되었다.

점차 도시화 되는 현대사회에 살고 있는 사람들. 현대화에 발
맞춰 걷고, 보고, 듣는 것에 익숙해진 우리들은 언제부터인가
낡고 헌것은 없어져야한다고 생각하고 있다. 이로 인해 재개
발이 강행되고 삶의 터전에서 강제로 쫓겨난 사람들은 어디
로 가야하는가?이번 계획안은 우리들의 '의식주' 중 '
주'를 대상으로 한 계획이다. 무분별한 재개발로 인해 살 곳
을 잃어버리는 사람들을 위해 어떻게 하면 조금이라도 더 개

선된 삶을 누릴 수 있을 지에 대하여 생각하게 되었다.

People live in a modern society that is becoming
urbanization. To keep up with the modernization,
we have become used to walking, seeing and lis-
tening, and started to think that old and worn out
things should be disappeared. Therefore where
should those people go who were forcibly evicted

from their houses due to the enforcement of the
redevelopment?
This proposal is targeting a shelter among our basic three necessities of lives. It has been thought
to find a way to provide an improved lives to
people who lost their places to live because of a
thoughtless redevelopment.

265

SITE ANAYLASIS

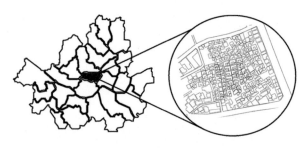

서울특별시 종로구 종로3가 낙원동
Nakwon-dong, Jongro 3 ga, Jongro-gu, Seoul

위치 : 서울 종로구
면적 : 2.96㎢
인구 : 1만 3723명(2008)
서울의 중심지답게 발전해 있는 종로구 그 화려한 도심지 속에 잘 살펴보지 않으면 발견하기 힘든 낙원동 쪽방 촌이 존재한다.

Location : Jongro-gu, Seoul
Size : 2.96㎢
Population : 13,723 (2008)
There are Nakwon-dong doss houses that are hard to be recognized in Jongro-gu where it is developed fancifully as a core of Seoul

서울특별시 세종로139번지 세종로 사거리에서 종로구 종로6가 79번지 동대문에 이르는 가로(街路)및 그 주변 동

Avenues and surrounding dongs between the Sejongro intersection in Sejongro 139, Seoul and Dongdaemoon in Jongro 6 ga 79, Jongro-gu.

우리네 이야기
Story about our selves

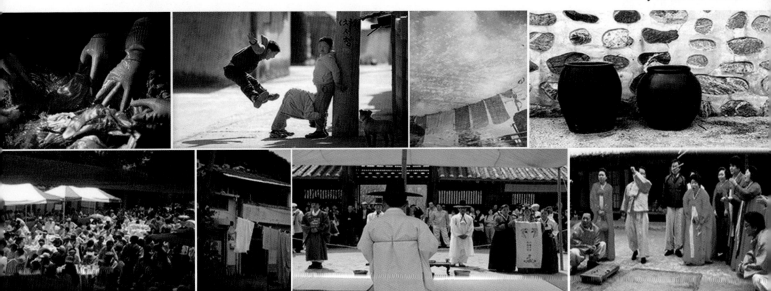

DESIGN CONCEPT

'젠가'란 나무 블록 탑의 맨 위층 블록을 제외한 나머지 층의 블록을 하나씩 빼서 다시 맨 위층에 쌓아 올리는 보드게임의 일종이다. '넣다, 짓다, 건설하다'라는 뜻을 가지고 있는 스와힐 리어이다.

'Zenga' is a type of board games that takes a block from anywhere except the very top and then places it on the very top. It is Swahili that has a meaning of 'put, build, construct.'

쌓다 – 여러 개의 물건을 겹겹이 포개어 얹어놓다.
물건을 차곡차곡 포개어 얹어서 구조물을 이룬다. 밑바탕을 닦아서 튼튼하게 마련하다.

Pile - piles up various things in layers.
By piling things up neatly in layers to make a structure. Build a foundation to have a firm structure.

끼우다 – 벌어진 사이에 무엇을 넣고 죄어서 빠지지 않게 하다. 무엇에 걸려 있도록 꿰거나 꽂다.

Insert - puts something in a gap to tighten so that it won't fall out. Thread or put to be hung by something. have a firm structure.

빼내다 – 박혀 있거나 끼워져 있는 것을 뽑아내다.
여럿 가운데서 필요한 것 혹은 불필요한 것만을 골라내다.

Extract - takes out embedded or inserted things. Sorts out needed or not needed things from various.

DESIGN PROCESS

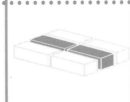

기존의 건물들은 일정한 형태의 모양을 가지고 있다. 이를 블록의 형태로 형상화 시켜 젠가의 쌓고 빼내는 방식을 이용한다.
Existing buildings have a uniform shape.
Figure these buildings as blocks and use piling and extracting methods of Zenga.

마당
벽체나 건물에 의해서 집 둘레에 부분적으로 또는 전체 적으로 둘러싸인 평평하게 닦아 놓은 빈 땅을 칭한다.

Madang
A flat empty lot that is built near a house and surrounded by dry walls or buildings either partially or entirely .

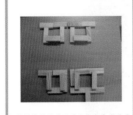
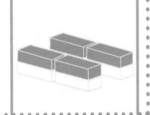

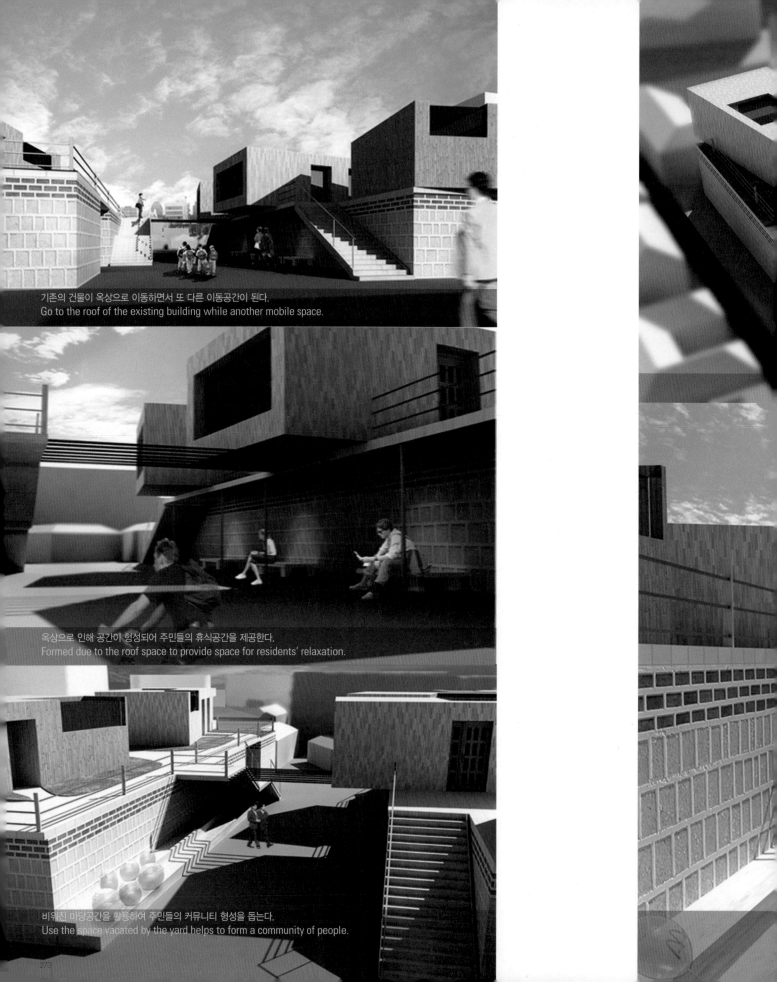

기존의 건물이 옥상으로 이동하면서 또 다른 이동공간이 된다.
Go to the roof of the existing building while another mobile space.

옥상으로 인해 공간이 형성되어 주민들의 휴식공간을 제공한다.
Formed due to the roof space to provide space for residents' relaxation.

비워진 마당공간을 활용하여 주민들의 커뮤니티 형성을 돕는다.
Use the space vacated by the yard helps to form a community of people.

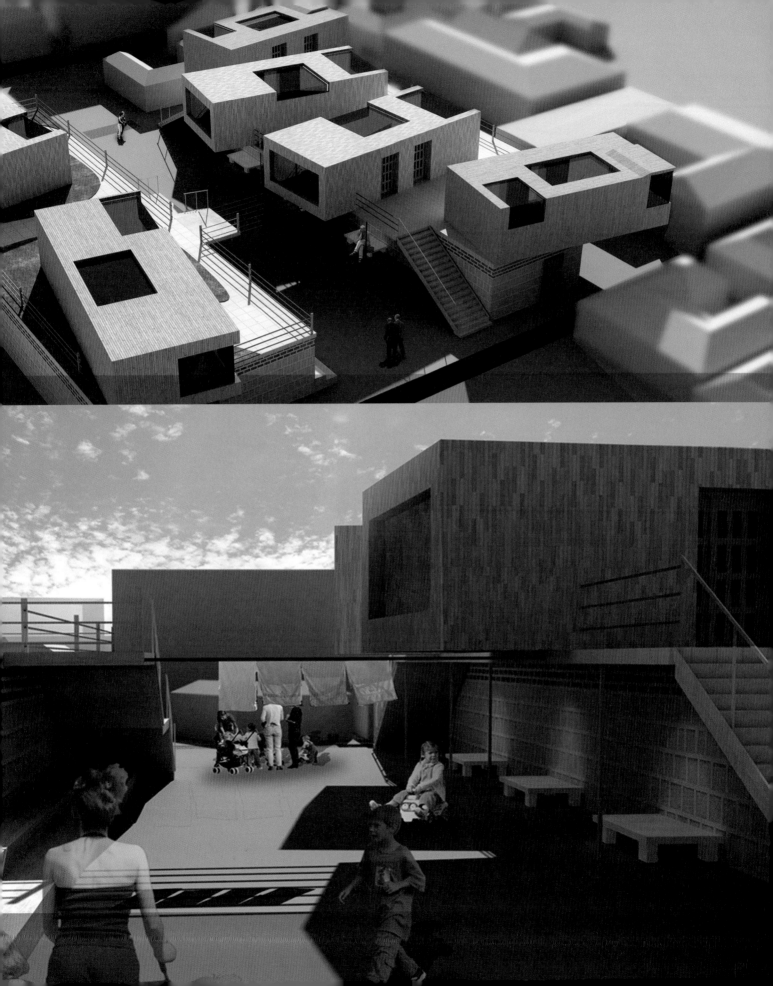

PLUG IN

공간과 공간을 연결하는 통로로 인한 주민들과의 소통이 원활한 옥상정원
The roof garden to communicate with residents well by the bridge connecting space and space.

BACKGROUND

사회적으로 문명, 기술이 날로 복잡해져 가는 현대 생활 환경 속에서, 사람들과의 관계는 점점 단절되어 가고 함께 하기 보다는 개인적인 형태로 변해가고 있다. 이러한 세태 속에서 오래된 동대문 아파트는 새로운 건축물이 즐비한 '서울' 이라는 대도시 중심에 위치하여 지역적으로 볼때 어울리지않고 소외되고 잊혀져 가고 있다. 주민들 또한 각박한 현대사회 속에서 서로의 소통보다는 단절된 모습을 보이고 있다. 이러한 주민들과의 단절된 상태. 더 나아가 지역적으로 소외되고 점차 잊혀져 쇠락하는 아파트를 옥상이라는 공간을 통해 함께 어울려 변해가는 구심점의 기회를 제공해 주려고 한다.

The relationship between people is cut off and changes to being personal rather than being together in the modern living circumstance which has more socially complicated technology and civilization. Under such social condition, old Dongdaemun Apartment is located in the center of large city, 'Seoul', where has numerous new buildings so that it has not been harmonized with the surrounding environment, excluded and forgotten. Residents also cut off neighbors rather than communicate with each other in the harsh modern society. The space of roof garden provides opportunity of the pivot to change such severed condition of residents, further, the apartment that has declined, excluded and forgotten locally.

DONGDAEMUN APARTMENT

CONDITION OF LOCATION

종로구 중심상권
도심재개발 산업
경복궁, 창덕궁, 종묘 등 문화유산 밀집
동대문 디자인 플라자 파크 조성
다양한 복합문화 유입

Core business area of Jongno-gu Urban redevelopment project
lots of cultural heritages such as Gyeong bok palace, Changdeok palace, and Jongmyo Shrine
Construction of Dongdaemun Design Plaza Park
inflow of various multi-cultures

SITE ANALYSIS

서울시 종로구 창신1동 328–17
5,300평
131가구
ㄷ'자 중정형/복도
328-17 Changsin- 1 dong,
Jongno-gu,
Seoul, Korea
5,300 pyeong
Number of household 131
courtyard / hallway of 'ㄷ' character type

\

TRANSFER

서울 시내 곳곳에 이어지는 20개의 버스노선 역세권 위치
지하철 이용의 편리함

20 bus routes travelling
throughout the city
located in transit area
convenience of subway use

CONCEPT

플러그가 콘센트를 찾아 꼽는 이미지에서 출발하였다. 플러그와 콘센트는 둘 다 없어서는 안될 중요한 구성 단위로 자체로서는 아무런 의미가 없으며, 서로 상호작용을 하지 않으면 존재가치가 없다. 즉 'PLUG IN' 이라는 단어 속에서 연결성, 접근성, 연계성이라는 개념을 부여하여 공간(콘센트)과 그것을 이어 주는 통로(플러그)가 만나게 되어, 주민들 사이에 커뮤니티를 활성화 시켜주는 매개체가 된다.

It starts from the image inserting a plug into the socket. The plug and socket are important and indispensable unit and they do not have meaning and worth per se without interaction.Therefore, the concepts of connectivity, accessibility, linkage are provided in word 'plug in' so that it becomes the medium to energize the communication between residents as space(socket) meets the passage (plug) to connect it.

PLUG IN SPACE

MOTIVE

도르래는 지레의 원리를 이용한 도구이다. 힘점, 받침점, 작용점 이 세가지 요소로 인하여 힘을 효율적으로 사용하는 것으로, 물체의 양에 비례해서 작용하는 힘도 변하는 가변적인 요소를 띄고 있다. 아파트 내부 양쪽을 마주 보는 중정 사이에 도르래가 위치하고 있다. 서로 마주보며 빨래를 너는 과정 속에 주민들 상호 관계가 형성 될 수 있다는 상징적인 의미를 띄고 있는데, 여기서 변화, 소통, 확장, 겹침 이라는 키워드를 뽑아 낼수 있다.

LOCATION

BUILDING

STATIONERY AREA

PUBLIC BUILDING

SUBWAY

ROAD

DESIGN KEYWORD & PROCESS

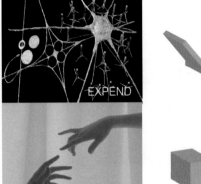

EXPEND

COMMUNITY

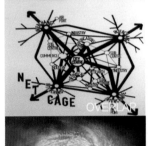

OVERLAP

CHANGES

TRAFFIC LINE

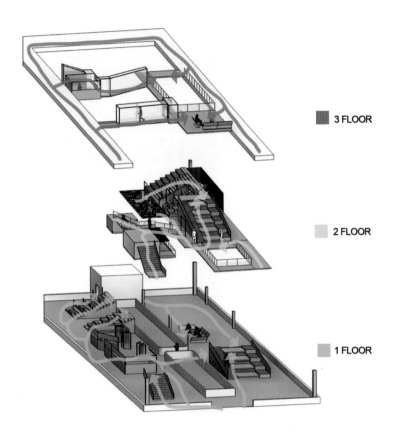

■ 3 FLOOR

■ 2 FLOOR

■ 1 FLOOR

ZONING

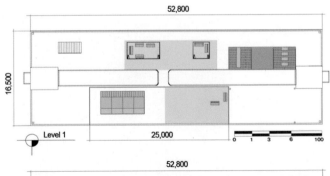

52,800

16,500

Level 1   25,000   0 1 3 6 100

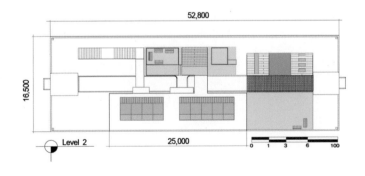

52,800

16,500

Level 2   25,000   0 1 3 6 100

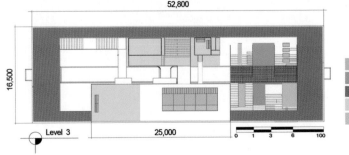

52,800

16,500

Level 3   25,000   0 1 3 6 100

■ 텃밭
■ 휴식공간
■ 산책로
■ 모임공간
■ 녹지

ELEVATION1

A. private and communal place to relax

B. place to dry laundry and communal place to relax

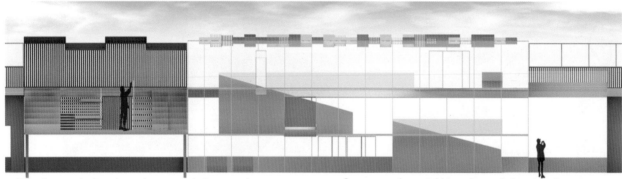

B-1. place to dry laundry and communal place to relax

C. communal vegetable garden and place to relax

A. 아래 공간은 벽의 이동으로 더 넓은 공간을 얻을 수 있음

private and communal place to relax- wider space can be earned by moving of wall for space below

B. 빨래를 너는 기둥이 계단 난간의 형태로 나올수 있음

place to dry laundry and communal place to relax- pillar to dry laundry can be formed as the banister of staircase

SPACE CONFIGURATION

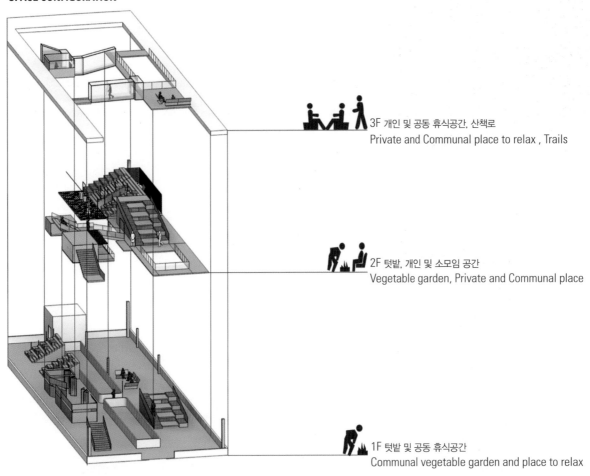

3F 개인 및 공동 휴식공간, 산책로
Private and Communal place to relax , Trails

2F 텃밭, 개인 및 소모임 공간
Vegetable garden, Private and Communal place

1F 텃밭 및 공동 휴식공간
Communal vegetable garden and place to relax

C. 주민들과 함께 식물을 재배하며 소통하는 공간
vegetable garden connected from the first floor to third floor - the space for residents to grow plants and communicate together

HOW TO USE ROOFTOP

옥상이라는 제한된 공간 속에 'plug in' 이라는 연결성을 통해서 변화를 나타내고자 했다. 동대문 아파트는 내부 중정이라는 특징 때문에 어찌보면 넓어보이지만 dead space가 존재한다는 약점이 있다. 이를 극복하기 위해 통로라는 연결성을 계단과 브릿지 형태로 나타냈다. 마치 커다란 멀티탭에 여러 가지 플러그를 꼽아놓은 듯한 형태로 '텃밭'이라는 거대한 커뮤니티 공간을 귀결하는 정점을 나타내고 있다.

It is presented the change by the connectivity of 'plug in' within the limited space, rooftop. The Dong-daemun apartment has a weakness that dead space exists because of internal courtyard despite of wide looking. To overcome it, the connectivity of passage represents as stair and bridge form. it represents the peak to conclude enormous community space of 'vegetable garden' like the form inserting several plugs into huge multi-tap.

기존의 이동하는 계단이 아닌 앉아서 쉴수 있는 벤치 개념이 포함 위, 아래로 조절이 가능한 빨래건조대

including the concept of bench which can sit by and take a rest rather than existing staircase to move drying rack to move up and down

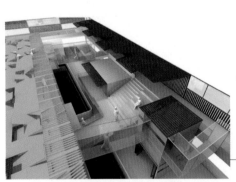

Rover integrating with roof railing

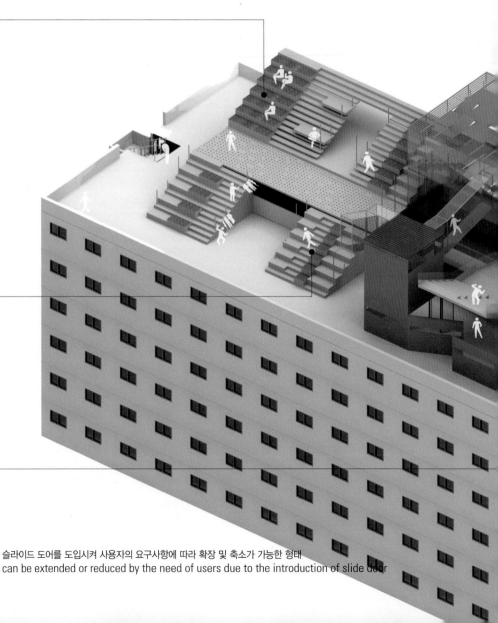

슬라이드 도어를 도입시켜 사용자의 요구사항에 따라 확장 및 축소가 가능한 형태
can be extended or reduced by the need of users due to the introduction of slide door

THE NEW INTERPRETATION OF 'ROOFTOP'

텃밭 전체가 유리로 되어 있어 태양열이실내로 유입되면서 식물재배에 알맞은 환경을제공하며 각 층마다 휴게공간이 있어 주민들모두가 함께 할 수 있는 공간 창출

providing suitable environment to grow plants by sun heats coming into the building through glass covering all vegetable gardens and making space to share with all residents due to the place to relax in each floor.

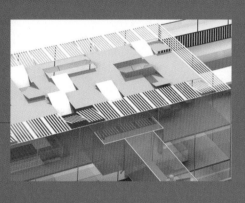

텃밭 공간 상부에 위치하는 solar cell로 태양의 높이에 따라 판이 상,하로 움직이며 빛을 흡수

absorbing the light by moving board up and down depending on the height of sun owing to the solar cell located in upper part of vegetable garden space

기존의 좁은 통로를 개선하기위한 것으로 각기 다른 높이에 공중 연결 통로

air passage at the different height for the improvement of existing narrow passage

텃밭 공간과 개인 및 공동 휴식 공간을 이어주는 통로 , 기존 옥상의 좁은 통로를 각기 다른 경사, 높이를 개선하면서 주민들의 소통을 원활해 질수 있음

communal vegetable garden and bridge to connect private and communal place to relax - residents can easily communicate each other because narrow bridge in the existing rooftop improves as bridges having different height and slope

각기 다른 공중 연결 통로 모습

Passage of different heights

개인 및 공동 휴식공간, 아래 공간은 벽의 이동

Private and communal place to relax- wider space can be earned by moving of wall for space below

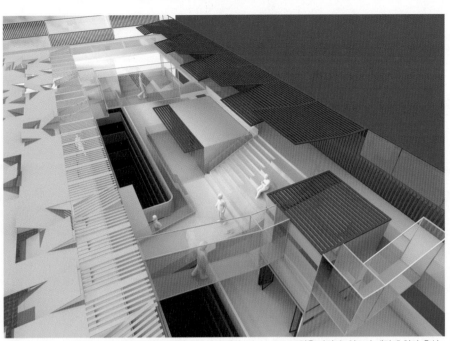

텃밭과 마주보는 개인 및 공동 휴식 공간, 밀폐된 공간에서 개인 또는 소규모 모임을 가질 수 있으며 계단에 앉아 휴식

Private and Communal place to relax facing communal vegetable garden private or small meeting can be held in the sealed space and taking a rest in sitting on the stair

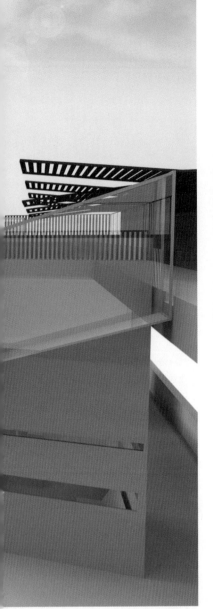

옥상 내부에서 본 텃밭 공간, 주민들과의 소통과 순환의 공간의 장

The side of communal vegetable garden residents can provide circulation and community space

공동 텃밭 입구 모습

the entrance of communal vegetable garden

樓 [ru:]
Foreign Tourist를 위한 Information Kiosk

SUBWAY ROOFTOP AREA

SUBWAY AREA

BACKGROUND

지하철 환기구는 지하철의 허파의 역할로 역사 내부의 공기를 항상 쾌적하게 유지하기 위한 장치이다. 그러나 환기구는 무방비 상태로 외부에 노출되어 지하철을 이용하는 시민의 건강을 위협하고 있다. 본 프로젝트의 목적은 도시경관을 해치고 사람들에게 불쾌감을 주는 환기구를 보다 쾌적하고 유용하게 사용할 수 있도록 하는데 있다. 지하철 공간의 옥상에 해당하는 지상(ground level)의 환기구 외부에 관광안내소를 계획한다. 그린 에너지가 전 세계적 이슈가 된 오늘날, 환기구 를 통해서 공기 중으로 버려지고 있는 폐열을 재활용한다. 환기구에서 배출되고 되는 공기의 온도는 섭씨 24도로 역사 내부의 온도와 같으며 지하철 운행시간 (05:30~24:50) 동안 지속적으로 배출되고 있다. 도심 속에 버려진 조각난 공간(환기구)위에 폐열을 이용하면서 동시에 여행정보를 제공할 수 있는 누각(kiosk)을 계획 한다.

The ventilation hole of the subway is the lung for the underground space that always keeps the air pleasant. However, the ventilation holes are exposed carelessly outside, threatening the health of subway users. This project aims to make the ventilation holes more agreeable and useful now that they spoil townscape and displease people. Tourist information centers are planned here to be built outside the ventilation holes on the ground level which equals to a rooftop for the subway space. Today green energy has become a global issue, in which sense the waste heat emitted in the air through the ventilation holes is recycled. The temperature of the air emitted from the ventilation holes is 24°C, which is same as that measured inside the subway stations. The waste heat keeps being discharged during the subway operating hours(05:30~24:50). A pavilion (kiosk) is designed here that can take advantage of the waste heat out of the scattered space (ventilation hole) dumped in the city center and provide tourist information.

CONCEPT

누각(樓閣) 다락루, 문설주 각(기둥) (층집 각)

[명사]
1. 사방을 바라볼 수 있도록 문과 벽이 없이 다락처럼 높이 지은 집
2. 이층이나 삼 층으로 지은 한옥
(누각의 사각형, 공간의 확장성, 외–내부 바람의 유입, 지붕에 사용되는 기와, 전통건축의 넉다운 구조, 서까래, 공포의 연속성)

Pavilion(樓閣) loft belvedere, jamb 각(pillar) (층집 각)

[Noun]
1. A house built high like a loft with neither doors nor walls to command a view in all directions.
2. 2- or 3-storied Korean traditional house(applied with the rectangular shape of a pavilion, spatial expansiveness, internal/external wind inflow, tiles on the rooftop, knock-down structure of a traditional architecture, rafter and continuity of 공포)

DESIGN PROCESS

누각의 사각형
한국 전통 건축물에 해당되는 누각은 기둥들이 사각형 형태 를 따라 세워져 있다.

A pavilion's rectangular shape : As a Korean traditional ar chitecture, a pavilion has pillars built following the rectangular shape.

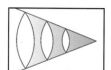

공간의 확장성
누각에서 외부를 바라 본 모습은 또 다른 시야의 확장감을 가져 다 준다.

Spatial expansiveness : Views seen from the pavilion give another expansiveness of a visual field.

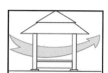

외내부 바람의 유입
벽이 없는 누각은 외부와 내부의 경계가 없어 바람의 유입이 쉽다.

Internal and external wind inflow : The pavilion having no walls allows easy inflow of the wind as it has no boundary between the exterior and the interior.

지붕에 사용되는 기와
기와지붕을 재료로 사용하는 전통건축의 특징을 사용한다.

Tiles used on the rooftop : As seen in traditional architecture using tiles on the rooftop.the exterior and the interior.

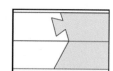

전통건축의 넉다운 구조
못이나 본드를 사용하는 것이 아니라 자연재료끼리의 접합방식을 사용

The knock-down structure of traditional architecture : Wood materials are combined not with nails or a bond but with natural materials.

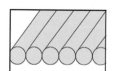

서까래와 공포의 연속성
같은 모양의 공포가 일정한 간격으로 놓여있는 연속성

Rafter The structure used for the rooftop like tiles the continuity of 'Gongpo' : Uniform 'Gongpos' for continuity by being laid at regular intervalsare applied to the design.

TARGET

TARGET : 외국인   SUBTARGET : 내국인

한국에 온 목적이 뭡니까?
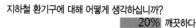
50% 관광
10% 쇼핑
10% 교육
30% 기타

지하철 환기구에 대해 어떻게 생각하십니까?
20% 깨끗하다
40% 더럽다
20% 보통이다
20% 기타

외국인들을 위한 시설 중 미흡한 부분은?
20% 편의시설
30% 흡연실
40% INFORMATION
10% 기타

SITE

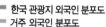

한국 관광지 외국인 분포도
거주 외국인 분포도

서울특별시 종로구
지하철 5호선 광화문역
지하철 3호선 안국역

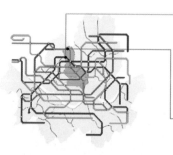

광화문 주변 상황 , 로케이션

안국역 주변 상황, 로케이션

지하철 노선과 중첩되는 지역

ROOFTOP DEFINITION

옥상은 건축물의 최상 층이며 열린 공간을 말한다.
A rooftop is the uppermost level of a building and an open space.

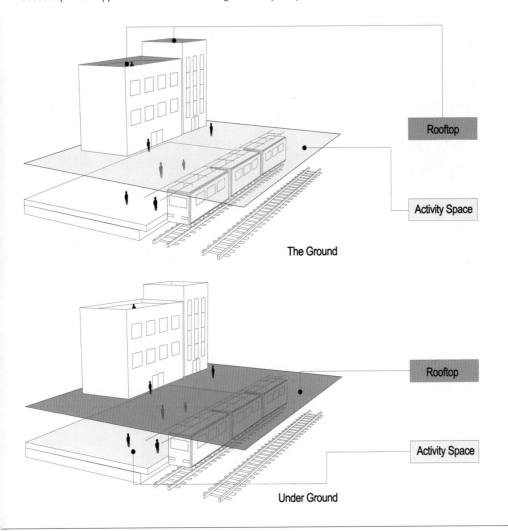

The Ground

Rooftop

Activity Space

Rooftop

Activity Space

Under Ground

우리가 활동하는 공간은 크게 지상과 지하고 나눠질 수 있으며 일반적인 지상에 세워진 건물의 맨 위층이 옥상이라 한다면 지하철의 옥상은 지상이라고 설정 할 수 있다. 본 프로젝트는 지하철 환기구에 초점을 맞추었다.

The space where we act and live is largely divided into two parts: underground and ground level. In general, the uppermost level of a building built on the ground is called a rooftop, in which sense the ground may be considered as the rooftop for the subway stations. The present project focuses on the ventilation holes of the subway stations.

STRUCTURE PROCESS

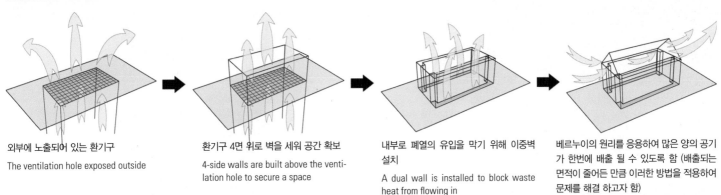

외부에 노출되어 있는 환기구

The ventilation hole exposed outside

환기구 4면 위로 벽을 세워 공간 확보

4-side walls are built above the ventilation hole to secure a space

내부로 폐열의 유입을 막기 위해 이중벽 설치

A dual wall is installed to block waste heat from flowing in

베르누이의 원리를 응용하여 많은 양의 공기가 한번에 배출 될 수 있도록 함 (배출되는 면적이 줄어든 만큼 이러한 방법을 적용하여 문제를 해결하고자 함)

Bernoulli's principle is applied to have lots of air emitted at once. (As the area of emission is reduced, this method is applied to solve the problem.)

내부에서 이루어지는 행위에 대한 것이다. 한쪽 벽에는 관광 및 각종 정보를 찾아볼 수 있는 터치스크린이 설치되어 있다. 반대편 벽면에는 서울 지도가 설치되어 있어 자신이 어디에 있는지, 목적지까지 어떻게 찾아가야 하는지에 대한 다양한 정보를 쉽고 빠르게 찾을 수 있다.

It is about actions performed inside. On one side of the wall, a touch screen is installed to help find tourist and other information. On the other side of the wall, the map of Seoul is installed to help people find quickly and easily where they are and how to get to their destinations.

벽면에 연출되온 모습은 누각이 갖고 있는 공간의 확장감을 표현하기 위해 거울반사를 사용하였다. 반사되는 피사체는 서까래로서 누각에서 쉽게 찾아볼 수 있다. 바닥에 설치되어 있는 기와는 천장의 거울에 반사되어 지붕이 있는 지붕 위를 걸어 다니는 착각을 준다.

A mirror is used to express the spatial expansiveness of the pavilion space reflected on the wall. The reflected subject is the rafter that can easily be seen in the pavilion. Rooftiles on the floor are reflected in the mirror on the ceiling, causing people to have an illusion that they walk on the rooftop.

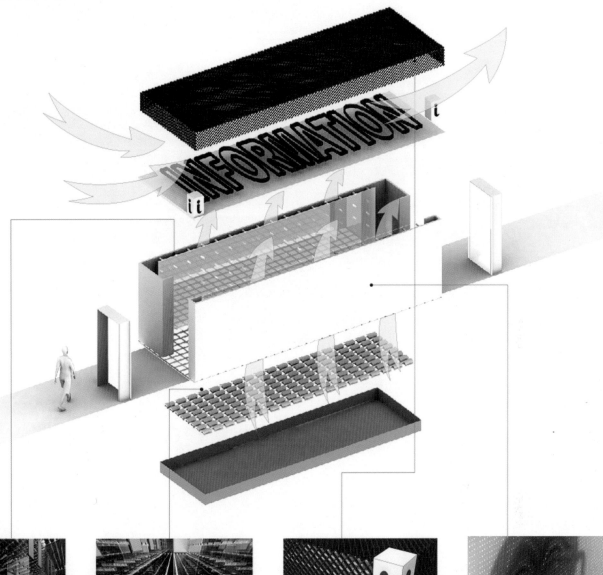

거울을 통해 서까래의 연속성과 공간의 확
장감을 함께 연출

The ventilation hole exposed outsideA
mirror is used to produce a continuity of
rafters and a spatial expansiveness

천장 거울을 사용하여 바닥의 기와를 극
적으로 보여줌

The ventilation hole exposed outsideA
mirror is used to produce a continuity of
rafters and a spatial expansiveness

넉다운으로 구성된 천장 구조물을 바람 길
로 사용

The ceiling is configured as a knock-down
structure.The structure is used as a path
for the wind

전체 벽면 LED 조명 설치

LED lights installed on the whole wall.

FLOOR PLAN

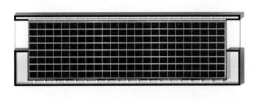

ELEVATION

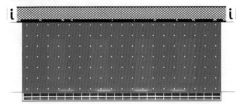

FACADE

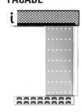

A tourist information center is built in a place where a ventilation hole is installed on a street in Seoul that cars and passengers pass by on the ground, while subway trains move underground.

LED ads on images of business or products

LED ads on images of business or products

LED ads on images of Korean tourist products

LED ads on images of Korean tourist products

The Remain
A Community of Their Own_Rooftop

REMAIN

이미 버려진 인천 시도의 폐교는 그 자체의 용도를 다하고 더 이상 이용 되지 않고 있다. 이러한 폐교는 어쩌면 이 곳의 주민들의 모습과도 같다. 폐교의 재생으로 인한 주민들에게는 새로운 문화와 커뮤니티의 장소로 남겨지게 되고 시도를 방문하게 되는 관광객들에게는 혐오감을 주는 건축물에서 새로운 관광의 장소로 재탄생하여 아름다운 추억을 머물고 갈 것이다.

이처럼 시도의 방문 하는 또는 거주하는 모든 이들에게 머물 수 있는 공간을 제공한다.

The ventilation hole exposed outsideA mirror is used to produce a continuity of rafters and a spatial expansiveness place to be left to try to visit a building that is repellent to tourists as a place of rebirth in the new tour to the beautiful memories are going to stay.

This attempt to visit or reside in any of these offers a place to stay for.

SITE

인천 옹진군 시도리 334-6
면적 6.92㎢
해안선길이 16.1km

Incheon Ongjingun sidori 334-6
Area 6.92 ㎢
16.1km coastline

BACKGROUND

1999년 12월 말 현재 165세대에 418명의 주민이 거주하고 있다. 인천광역시에서 북서쪽으로 14km, 강화도에서 남쪽으로 5km 떨어진 지점에 있으며, 최고점은 구봉산(九峰山:178.4m)이다. 지명은 이곳에 사는 주민들이 성실하고 순박하다는 뜻에서 유래되었다고 하며, 진짜 소금을 생산하는 곳이라 하여 진염(眞鹽)이라고도 한다. 주변의 유명 관광명소는 유명한 한국드라마들 중 "풀하우스", "슬픈연가", "연인"," 시간" 을 촬영한 촬영지들이 있고 여러가지 조각상들이 있는 " 배미꾸미조각공원" 이 있다.

At the end of December 1999 of 418 165 inhabitants for generations are living. Incheon Metro poli tan City, northwest of the 14km, 5km south of the strengthening of points away, and is the culmination gubongsan. Named for the people living here means that comes from a sincere and naive, and really that's because the production of salt is a jinyeomyiragodo. South Korea's most famous tourist spots around the drama of the famous "Full House", "Sad Love Story," "Lover", "time" taken by various sculptures and the Location of "baemi Ornament Sculpture Park" is.

노인들은 사회적약자로 인식되어 있다. 몸과 머리가 점차 노화되어 감을 전제로 이러한 인식을 우리가 방도를 제시하려 한다. 인천의 끝자락, 도심지에서 떨어져 있는 바다위의 작은섬 인천신도마을을 정하였다. 신도를 가기위해선 교통수단이 배한척뿐이다. 이곳의 주민들은 대부분이 노인들로 구성되어 있고, 불편한 교통수단 때문에 외지사람들의 방문은 그리 많지 않다. 이런 그들의 외로움을 덜어주기위한 커뮤니티 공간을 제시 또한 노인들의 노화를 다시 재생시켜주는 재생 · 재활의 공간 제시를 할 것이다.

Elderly people are regarded as social outcasts. Sense of an increasingly aging body and head is no way presupposes the rec ognition that we are trying to present. The edge of Incheon, Incheon jakeunseom of downtown away from the sea, the village was determined believers. Shinto is a boat transportation to the top instantly. Its inhabitants are mostly composed of old, inconvenient transportation out here because there are not many people visit. They relieve the loneliness of such a community space, presented a re-aging of the elderly, rehabilitation Renewing your presentation will have the space.

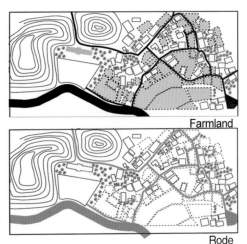

Farmland

Living space

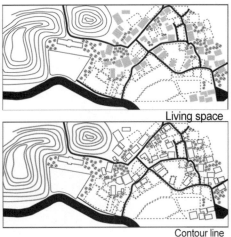

Rode

Contour line

PROBLEM

커뮤니티
공간의 부재

APPROPACH.1

여가 시간의 활용장소 제시 + 주민들만의 커뮤니티 장소 제시
= 노인들에게 즐거움의 장소 제공

Spare time, take advantage of the present + residents own community, a place where provided = provided a place of joy to the elderly

APPROPACH.2

신도의 문화 축제 공간 + 학교라는 공간에서의
교육 & 놀이 공간 제시
= 노인들을 위한 재생ㆍ재활이 필요한 교육&문화공간 제공

Followers of the school's cultural festival space + space provided in the Education & play = play for the elderly, rehabilitation, providing the necessary education & cultural space

→ [커뮤니티 공간의 제시]

→ [community space]

Complex

Resident

Closing school

Community recycle

→ [교육장소성의 부활]

→ [Resurrection of training places]

Refresh

Education

Watch

Community recycle

CONCEPT

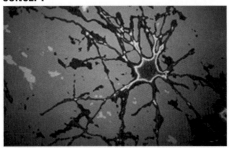

옥상에서의 보호와 커뮤니티의 공간을 형성해주는 것이다.뿐만 아니라 membrance는 그 형태적 특징으로서 뻗어나감과 집중되는 뉴런과도 같은 형태를 띠고 있는데 이러한 형태적 특징에서 옥상의 집결하여 사람들은 외로움이라는 결핍을 해결 해주고 다른 공간으로의 연결성을 이루워 줄 수 있는 공간적 특성을 가질 수 있게 된다.때문의 단순한 폐교로서 아무런 의미가 없는 공간에서 새로운 장소성을 부여한 문화 커뮤니티 복합 장소를 제안 한다.

Membrane

조직을 보호하는 얇은 막의 형성으로서 상징성은 그 지역의 주민들인 노인층의 인구를 감싼다는 의미를 가지고 있다.

Formation of a thin membrane to protect the organization as a symbolism deulin elderly residents of the region's population has the means to wrap.

In the roof space protection and community ties is. Membrance as well as the morphological characteristics of neurons with ppeoteona sense and is concentrated in the same form, which culminated in the morphological characteristics of these people gathered on the rooftops of loneliness, lack of resolve by giving the other space that could yiruwo spatial connectivity will be able to have attributes.

No meaning because there is no simple closed space as a sense of place given the new cultural complex where the community is proposed.

KEYWORD & PROCESS

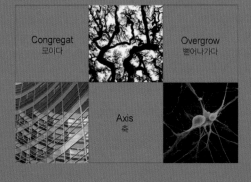

Congregate (모이다)

옥상으로서의 모임 즉, 커뮤니티 공간의 형성을 위한 집결로서 Membrane의 형태적 특징에서 집중되는 핵의 공간이 옥상으로 상징화 된다.

Membrane of the morphological characteristics That meeting, as the roof, making a space for community gatherings as Membrane concentrate on the morphological characteristics of space on the roof of the nucleus is symbolized.

Axis (축)

앞서 설명된 뻗어나감의 형태적 특징에서 bridge의 형태가 유기적 형태가 아닌 직선적인 축의 구성을 이루어 노인들에게 조금 더 명확한 산책로의 제시를 하게 되고 파생되어 각각의 축을 이용한 보다 편안한 산책로의 슬로프를 제시하고 형성된 슬로프에서 편안한 눈높이의 관람석을 제공하며 공간의 분할을 확실하게 만들어준다.

Ppeoteona sense described above in the form of bridge in the morphological characteristics of the organic forms made to configure a non-linear axis for the elderly to little more clear presentation of the trails and is derived using each of the axis to present a more comfortable walking with a slope of formed on the slopes of the seat provides a comfortable eye level, and certainly makes the division of space.

PROGRAM

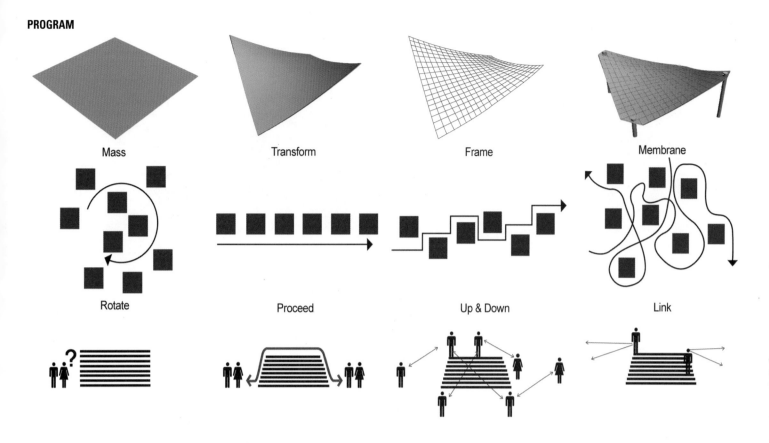

Mass

Transform

Frame

Membrane

Rotate

Proceed

Up & Down

Link

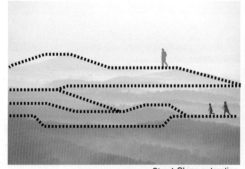

Step1 Slope extraction

Step2 Mountain

Step3 Line extraction

Overgrow (뻗어나가다)

앞서 형태적 특징 중 집중 집결의 의미를 가진 형태에서 반대로 다시 뻗어나간 다는 느낌을 실체화 시켜 구연시킨다. 이러한 뻗어나가감에서 옥상 뿐만 아니라 학교의 전반적인 bridge의 형성으로 다른 공간으로의 연결성을 구성해주고, 단순한 bridge의 역활 뿐만 아니라 지역주민인 노인층의 산책로의 형성과 학교의 재탄생에 의한 축제 문화를 관람할 수 있는 관람석의 형태를 이룬다.

Gathering of the morphological characteristics previously concentrated form with the meaning of the feeling that stretches back to the opposition by the materialization makes oral. In this sense ppeoteona the roof of the school, as well as the formation of the overall bridge to another room to do the configuration of the connectivity, as well as the role of a simple bridge of the elderly jiyeokjuminin the formation of trails by the rebirth of the school to watch the festival culture constitutes a form of the bleachers.

Cover (감싸다)

Membrane의 형태적 특징 중 가장 크게 두드러 지는 면으로서, 공간의 형태를 넘어 그 공간 자체를 이용하는 주민에게 보호의 느낌을 주는 천막을 형성하여 중점이 되는 옥상에서 가장 편안한 느낌과 보호받는 다는 느낌을 받을 수 있게 된다.

The most significant morphological characteristics of Membrane myeoneuroseo are apparent, the shape of the space beyond the space by itself gives the impression of protecting the residents formed a tent on the roof of the emphasis that is most comfortable feel and can feel protected are able.

KEYWORD& PROCESS

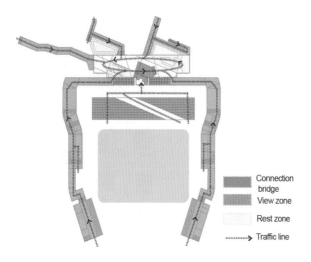

- Connection bridge
- View zone
- Rest zone
- ·········> Traffic line

SPACE ZONING & PROCESS

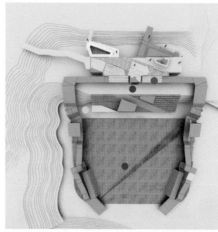

- ● Form 1 Roof top
  커뮤니티의 공간, 관람의 공간
  Community space, Watch zone

- ○ Form 2 Bridge
  산책로의 공간, 관람의 공간
  Walking space, Watch zone

- ● Form 3 interior
  교육의 공간
  Education space

- ● Form 4 Stadium
  문화축제 공간
  Culture festival space

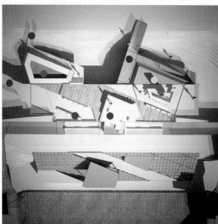

ROOF TOP ZONING

- ● Zone 1 관람석 (Watch zone)
- ○ Zone 2 카페 (Cafe)
- ● Zone 3 휴식공간 (Rest Space)
- ● Zone 4 놀이공간 (Game Space)
- ● Zone 5 연결다리 (Connetion Bidge)

Stand and Bridge
Bridge as a connection to the roof of the space is not a simple, fast trails and Research, in a cultural festival grounds in the bleachers to watch the space.

Education Space
Interior space as a space of education and training received by the computer and the resulting regenerated and cultural life.

Entry into Roof Top (Walking)
Section of the right to access state space walks,and a panoramic view overlooking the village and stadium tour of the cultural festivals can be.

Entrance
Exodus of the entrance room of the school entrance to the indoor playground at the role and cultural space at the same time to see it when you stand, you can watch the space.

Education Space
Interior space of another educational theories are taught using presentation place.

RoofTop Space
The roof space was a concept that best reflects the tent structure is protected by feel, communicate, and the space is an apex of the whole school walking space.

SITE PANORAMA

서울특별시 구로구 신도림동

현재 대상 사이트는 주위의 재개발속에서 얼마 남지 않은 낡고, 허름한 오래된 낙후 지역
이다. 큰 길 쪽으로는 철강소가 있고, 안쪽으로는 거주지역이 남아있는 개발이 필요한 지
역이다. 현재 재개발 계획에 포함된 지역이지만 구체적인 계획안이 제시된 상태가 아니기
때문에 재개발이 진행되고 있지 않은 상황이다.

Sindorim Dong, Guro GU, Seoul

This area is one of few shabby and backward parts surrounding redevelopment
projects. There are steel works on the main street and , behind the street, there
is the residential area, which need to be redeveloped. Those areas are planed
to redevelop, but plans have not been definite, so the areas are not on redevelopment right now.

別墅庭園 별서정원
Sharing the Garden

자연에 귀의하여 유유자적한 생활을 즐기려 만든 정원
To enjoy a comfortable life in nature, created by the ear garden

BACKGROUND

서울의 대표적 공업 단지 중 하나였던 신도림동 이제대부분의 지역이주거, 상업 단지로 변화되었고 , 현재 얼마 남지 않은 상업 단지 들도재개발 계획에 포함된다. 우리는 옛 것을 헐며 새로운 도시를 만드는 기존의 재개발 방법에서 벗어나 옥상 공간을 활용한 과거, 현재, 미래가 함께 공존해가는 방법을 제시한다.

Sindorim dong used to be one of the typical industrial city. However, It has been changed. Most of the area became residential and commercial area, and the rest of it are on redevelopment projects, We suggest the way utilizes the space of rooftops, which coexist past, present and future instead of the way just pull traditional buildings down.

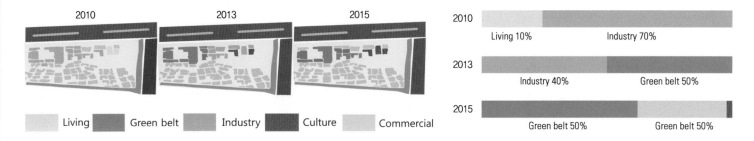

2010	2013	2015

Living | Green belt | Industry | Culture | Commercial

2010	Living 10%	Industry 70%
2013	Industry 40%	Green belt 50%
2015	Green belt 50%	Green belt 50%

DESIGN CONCEPT

"소쇄원"

조선시대 대표적 정원인 소쇄원은 자연 공간으로 휴식과 교류의 기능을 가지고 있다. 시간의 흐름에 따라 변화하는 자연을 통하여 시간의 흐름에 따른 공간의 변화를 느끼게 된다.

"Soshawon"

Soshawon, the representative garden of the Joseon Dynasty, used to be a nature place to relax and communicate. As time went by, people could feel changing the natural environment and space at the same time.

KEYWORD

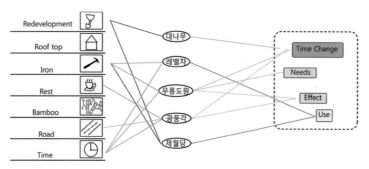

소쇄원의 공간구성 요소중 중요한 '휴식과 교류, 철공소 옥상위에 길이라는 요소와 대나무라는 요소를 끌어들인다. 옥상 공간의 가장자리를 대나무로 두르고 자연적 요소를 도입하여 정원에 온 것 처럼 느낄 수 있게 한다. 또한 길을 만듦으로써 생기는 자연스러운 공간 위에 사람 간의 교류 할 수 있는 몇 가지의 공간들을 만들어 별서정원 이라는 자연 정원을 연출하려 한다.

The most important things in space composition factors are relaxing and exchanging. On the rooftops of steel works, the elements, ways and bamboos are being brought. The rooftop is surrounded by bamboos and decorated by natural factors in order to make people feel they are in a garden. By building ways and spaces related to the ways, people could exchange things and communicate each other in this artistic and natural garden.

DESIGN PROCESS

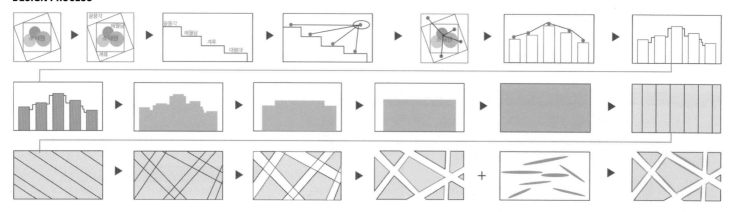

NEEDS

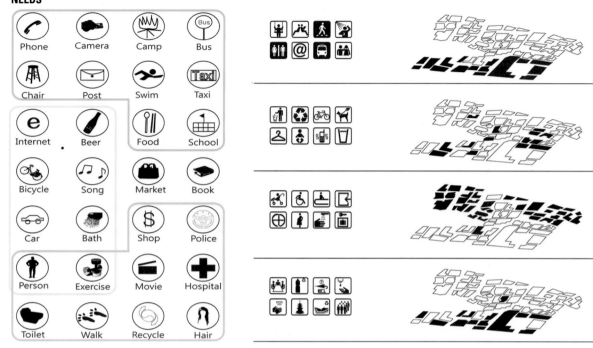

Phone	Camera	Camp	Bus
Chair	Post	Swim	Taxi
Internet	Beer	Food	School
Bicycle	Song	Market	Book
Car	Bath	Shop	Police
Person	Exercise	Movie	Hospital
Toilet	Walk	Recycle	Hair

Change 1

Collect 2

Connect 3

harmony 4

Required Function A ~ F group

A- group
B- group
C- group
D- group
E- group
F- group

소쇄원의 공간 구성을 신도림의 낙후된 철공소 위에 시간의 흐름에 따라 재현해 낸다.
−2010년 옥상 위에 도시 속 정원이라는 공간을 제시한다.
−2012년 옥상 위에 정원 공간이 들어선다.
−2013년 재개발 취지에 맞춰 철공소와 주거공간이 철거 된다.
−2014년 기존의 철공소 공간에 상업공간이 완성된다.
−2015년 전시공간과 영상공간, 오피스공간이 완성된다.

The Space composition of Soshaewon is reproduced on the backward steel works in Sindorim with time,
-By the year 2010, suggesting the garden in the city on rooftops
-By the year 2012, having been built the garden on rooftops
-By the year 2013, demolishing the steel works and residential areas according to the redevelopment.
-By the year 2014, completing commercial area on the place used to be steel works.
-By the year 2015, having been built exhibition and projection space, additionally, office space behind the spaces,

SPACE PROGRAMMING

신도림 철공소는 현재 재개발 하려는 시기와 맞물려 있다. 재개발이라는 취지에 어울려 우리는 별서정원이라는 조선 시대 최고의 휴식공간에서의 자연적인 요소와 내부 공간 구성을 통해 낙후되고 발전 가능성 없는 기존의 철공소 옥상위에서의 발전가능성을 제시한다.

The steel works in sindorim is going to be redeveloped, I suggest enormous potential for development on the rooftops of the steel works, which don't seem to have potential for development by bringing natural elements and interior space composition from the best artistic garden of the Joseon Dynasty.

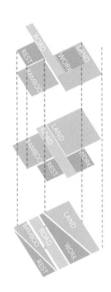

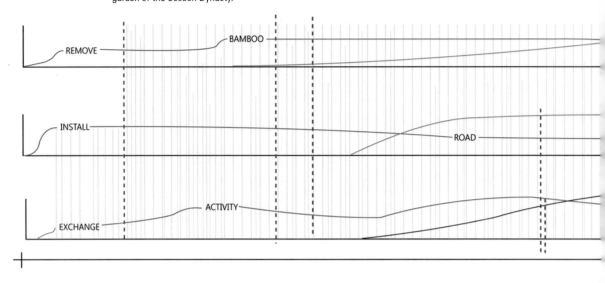

2011 PLAN   2012 PLAN   2013 PLAN

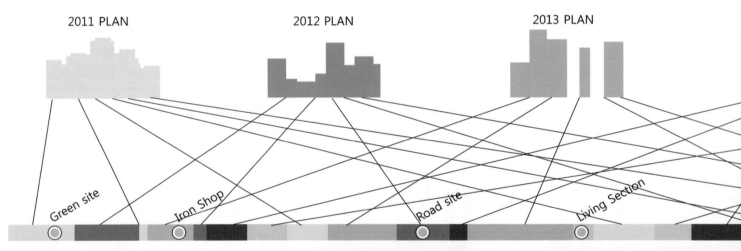

Green site   Iron Shop   Road site   Living Section

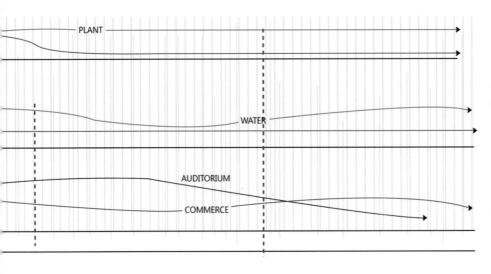

PLANT

WATER

AUDITORIUM

COMMERCE

4 PLAN   2015 PLAN

et Place   E.T.C section   Development   Future Section

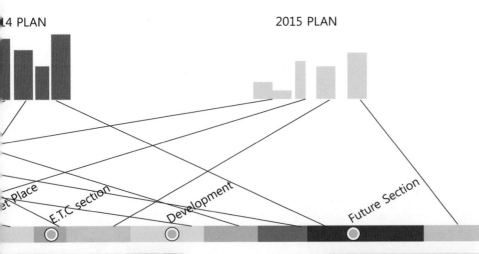

Permeate
economy + culture + communication = vegetable garden

Self-sufficiency _ ROOFTOP
버려지고 못 쓰는 공간을 스스로 만들어가는 텃밭과 가든. 영국이나 가까운 일본에서의 `OPEN GARDEN]의 개념을 더하여 사람과 사람의 소통과 공동체를 이루는, 그리고 함께 만들어 나가는 곳, 즉 나만의 밭, 우리의 옥상, 모두가 하나 가 될 수 있는 어우름의 공간을 제안.

It comes to throw away, it makes the space where it does not write oneself and it goes with the kitchen garden. Great Britain or it becomes worse `OPEN GARDEN' concepts from near Japan and it accomplishes and a possibility all becoming one and the place to make together, the namely only my field and our roof, there is the space where it does together it proposes being understood and the community of the person and the person.

BACKGROUND

옥상 공간 + 사회적 약자 라는 제약을 안고 사회적 약자를 판단하는 기준이 무엇이며 그들이 원하는 것은 무엇일까? 그리고 지금 이 작업을 하는나. 우리 자신들도 사회적 약자의 바운더리 안에 포함되어 있는 일부 이지 않을까? 그럼 우리(=사회적 약자)가 진정으로 원하고 필요한 공간은 무엇일까? 수년간 한번 쳐다보지도 올라가 보지도 않는 버려진 곳 옥상. 그곳을 우리 삶의 터전의 일부이자 일상으로 디자인하려고 한다.

Weak person roof spatial + `social' the standard which weak person social holds the restriction which is and judges will be what and they wanting will be what. And it will work now and or, also our themselves will not be fortune of weak person social compared to one intelligence which is included inside? Like that us (weak person = social) in true feelings to want the space where it is necessary will be what ?

SITE

서울시 중구 회현동1가
147-23 회현시민아파트

Seoul Joong Gu Hoe hyeon - dong 1 147-23 Heo hyeon civil apartments

SITE ANALYSIS

위로는 남산공원 밑으로는 남대문시장과 오른편으로 명동. 왼편으로는 시청이 위치한 서울의 핵심이라고 볼 수 있는 곳에 위치
할렘으로 여기지는 이곳은 영화나 방송에 범죄자들의 아지트나 불우한 환경으로 자주 등장시키곤 한다. 또한 이곳을 지나가는 사람들과 호기심에 들리는 사람들은 '그리움', '추억' 이라는 말로 포장을 하지만 현제 거주하는 주민들은 위에 언급했던 사회적인 부분이나 외부인들의 시선, 할렘가라는 낙인으로 인해 마음을 닫았다. 외부인들의 출입을 단절시키고 각종 경고문들이 붙어있으며 이곳을 지나가는 타인들을 적대시한다. 반면 거주민들 끼리의 공동체는 아파트의 문화와 달리 시골의 정겨운 풍경 을 보듯 가족과 같은 화합과 단결을 이룬다.

HISTORY

한국의 수도 서울은 아파트 공화국으로 알려져 있고 자산의 가장 큰 부분을 차지하고 있다. 좋은 아파트에 살아야 잘 산다는 공식이 성립되는, 매년 부동산 값의 변동과 아파트의 재개발 등의 사회적 문제가 끊이질 않는 곳. 그러한 회현 시민 아파트는 불과 40년이란 세월을 훌쩍 넘긴 아파트이다. 최장수 아파트인 이곳은 서울시와 세대주간의 대지소유권의 갈등과 대립으로 인해 10여 년간 재개발이 미뤄져왔고 매매도 불가능한 상태이다. 앞으로도 재개발이 언제 이뤄질지 모르는 낙후된 곳이다.

Capital Seoul of the Republic of Korea is become known as the apartment republic world-wide and widely the actually property it kicks major portion it is doing. It lives in the good apartment and lives the formality is formed well, The place where the social problem of redevelopement etc. of fluctuation and the apartment of every year real estate price is many. The such sliced raw fish present civil apartment only 40 the apartment which passes over the time when is with a jump is. The longest possibility apartment person the place is caused by discord and opposition of the earth right of ownership of Seoul and generation daytime and about 10 for year the redevelopement comes to put off, it comes and it is a condition whose also selling and buying is impossible. The redevelopement when coming to accomplish in future, it does not know and it is a place which falls behind.

RESIDENCE AREA

CULTURE AREA   TARGET AREA   GREEN AREA

버려진 공간   공간의 협소   외부와의 단절   +   주민 들의 끈끈함   =

Natural   Econonmy   Human

Comfort at the side which rises with a South Gate market with the Namsan park lower part it rumbles, with the left side as core of Seoul where watching is located it is located in the place which is the possibility of seeing.
With the RAM which it will do does not think this place in movie or broadcasting the agitation point of the criminals or appearance at the time of height boiled frequently with the environment which is unfortunate. Also the people who passes by this place and the people who is audible in curiosity `yearnings', `reminiscences' packing but the present me the residents which dwell gaze of the parts which are social it refers a above or the outsiders, the RAM which it will do go at the end which is with the brand to be caused by, it closed a mind. It discontinues the en trance and exit of the outsiders and the various warning messages are sticking and it is hostile to the others who pass by this place. The other side dwelling the community of the field among ourselves where it pushes culture of the apartment and the differently the country sees affectionate scenery with the family same accomplishes a harmony and unity.

ISSUE

유권분쟁의 이유로 아파트 매매가 이뤄지지 못해 현재 입주자가 없는 버려진 세대가 존재 입주자 사이의 단결과 협심이 높고 공동체 행사를 자주 벌임. 커뮤니티 공간의 협소, 노인정 및 놀이터 등의 공간이 일체 존재하지 않음.

Type apartment sale price accomplishing support not to be able there is not the present my moving in own house with moving in person and Cooperation opens a moving in person and Cooperation opens a community event highly frequently. community space narrow. The old person affection and space of playing etc. do not exist all.

SOLUTION

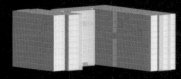

STEP 1. Normal

입주자가 없는 삭막한 죽은공간

Moving in own houses the gruel which is dreary space

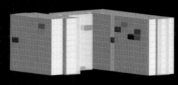

STEP 2. Dead Space

삭막한 죽은 공간을 새로운 공간으로 재탄생

The gruel which is dreary in the space where it is new a space re-birth

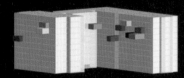

STEP 3. Subtraction

죽은 공간을 빼내어 Void & Soild의 긴장감을 부여하여 비워진 공간은 공간의 연결성과 쉼 터로 재탄생

It draws out the space where it dies and, Void & Solid feelings of tension and the space where to empty it gives it comes public affairs

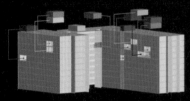

STEP 4. Raise

빼내어진 죽은 공간은 옥상에서 Vegetable Garden 으로 재탄생

It pulls out and the gruel space from the roof with Vegetable Garden re-birth

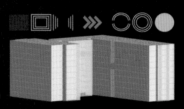

STEP 5. Change

빼내어진 공간을 사각 모듈에서 원 모듈로 변형하여 공간에 적용

To pull out and from square module to change a gentle space at circle module, in space applicaation

APPLICATION

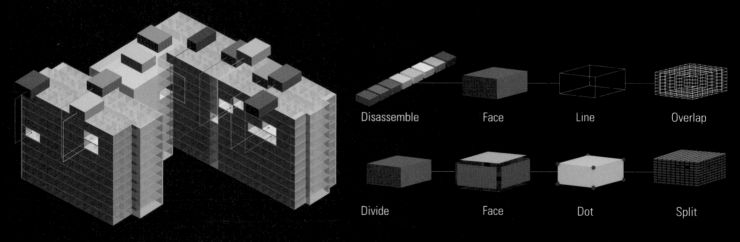

Disassemble Face Line Overlap

Divide Face Dot Split

CONCEPT

"WAVE INTERSECTION"

"wave intersection" 은 wave가 가지고 있는 특성에서 공명, 중첩, 파장과 원의 선을 이용하여 공간을 표현하는 집합체이다. 이러한 특성들이 공간과 공간을 만드는 요소들이다. 각 공간과 개인의 독립성과 개인적 이기주의를 탈피하고 서로 화합하여 하나가 되는 문화 운동 만남 경제의 울림으로 공동체의 "vegetable garden" 으로 재탄생된다.

"wave intersection" are fair from the quality which wave is having it is an aggregate which a space and, to use the line of reiteration, wavelength and circle expresses. They are elements where like this qualities make a space and a space. Each space and independence and personal egoism of the individual it exuviates and it harmonizes each other and the cultural motion where one becomes the economy meeting with ringing, "with vegetable garden" of the communities it re-is born.

KEYWORD & DESIGN PROCESS

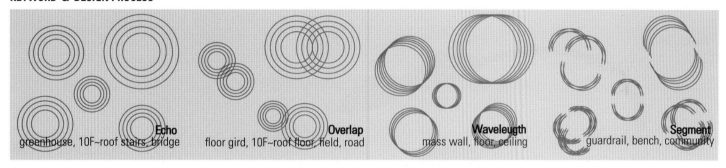

Echo
greenhouse, 10F~roof stairs, bridge

Overlap
floor gird, 10F~roof floor, field, road

Waveleugth
mass wall, floor, ceiling

Segment
guardrail, bench, community

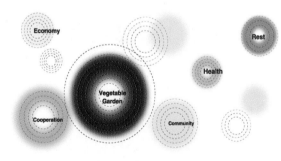

SUGGESTION

동심원은 울림으로 퍼지고 그 각각의 울림은 경제, 문화, 운동, 소통, 화합의 흐름이 되어 동심원들의 집합체로써 vegetable garden으로 탄생된다.

The concentric circle spreads out and with ringing, the respectively it rings and economy, culture, motion and being understood, harmony flowing, with vegetable garden it becomes and with the aggregate of concentric circles it is born.

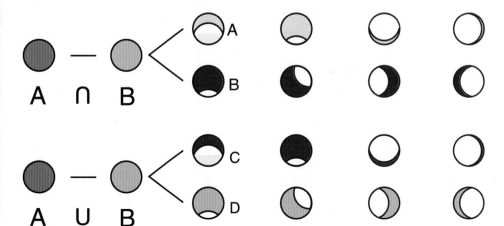

MASS PROCESS

wave의 특성중 중첩이라는 특성을 가지고 공간을 연출하였다. 중첩에서 합집합, 교집합이라는 요소를 가지고 다양한 형태의 조합과 변형으로 공간을 제시한다.

It had the quality which is reiterations in wave qualities and it produced a space. It has the element which is unionsand intersection from reiteration and a space with the union and variation of the form which is various it presents.

MASS PROCESS

죽은 공간을 원이라는 모듈로 변형시켜 wave intersection이라는 컨셉을 가지고 용도에 맞는 분할 조합 배치를 통해 다양한 공간을 연출한다.

It changes the space where it dies at the module which is a circle and they are wave intersection and concept it has and the division union arrangement which hits to a use it leads and the space where it is various it produces.

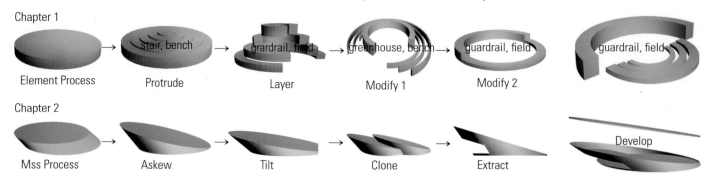

Chapter 1

Element Process → Protrude → Layer → Modify 1 → Modify 2

stair, bench / grardrail, field / greenhouse, bench / guardrail, field / guardrail, field

Chapter 2

Mss Process → Askew → Tilt → Clone → Extract

Develop

ISOMETRIC

Bench & Field, Sola cell
Keyword : Overlap, Echo, Waveleugth

Greenhouse, Vegetable field
Keyword : Overlap, Echo, Segment

Guardrail
Keyword : Overlap, Segment

Roof Floor
Keyword: All

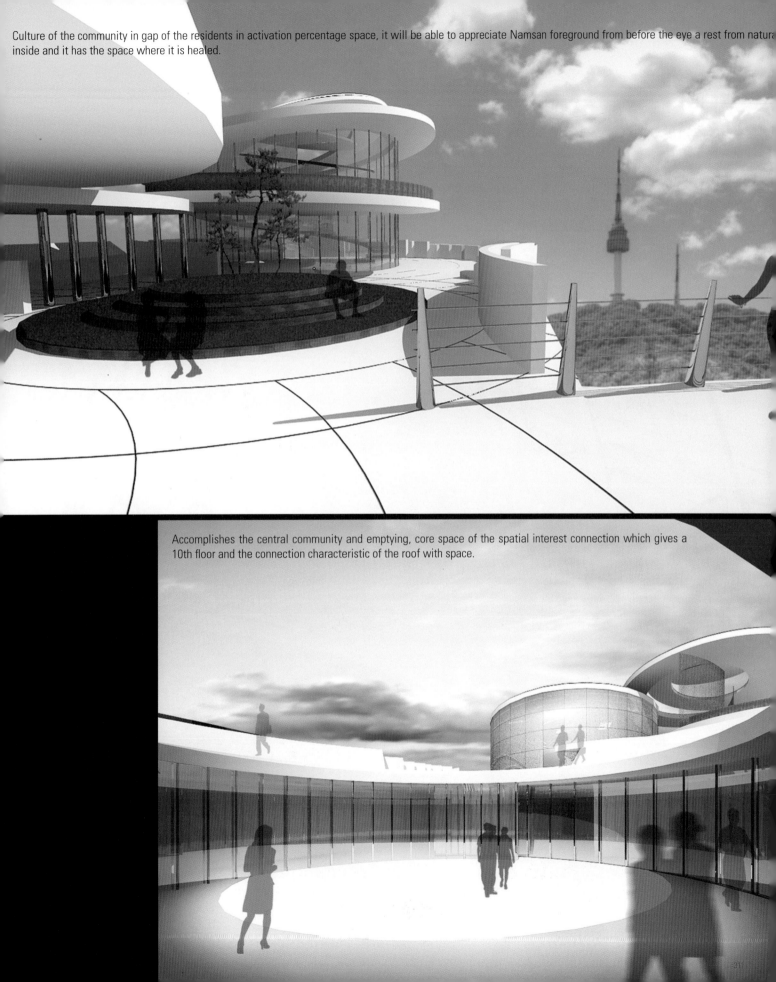

Culture of the community in gap of the residents in activation percentage space, it will be able to appreciate Namsan foreground from before the eye a rest from natura inside and it has the space where it is healed.

Accomplishes the central community and emptying, core space of the spatial interest connection which gives a 10th floor and the connection characteristic of the roof with space.

Features of the leg which has the function of the connection result observation platform of the roof and the roof.

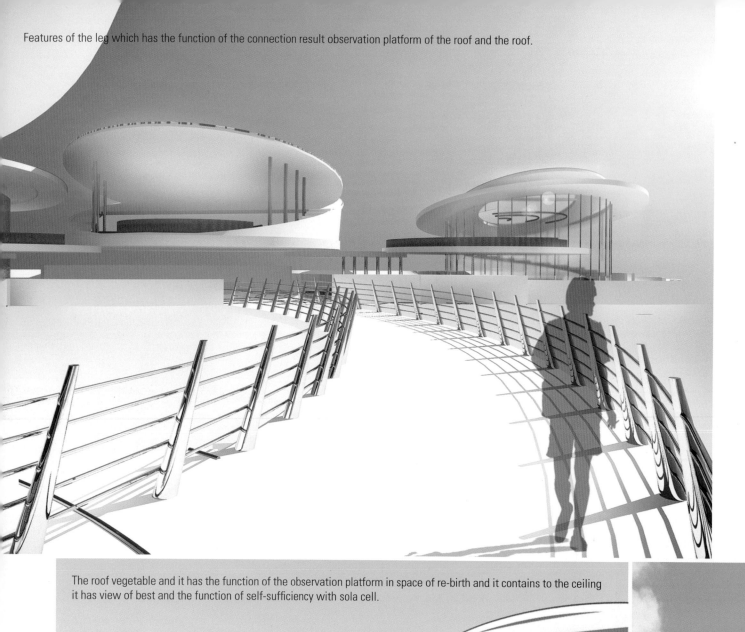

The roof vegetable and it has the function of the observation platform in space of re-birth and it contains to the ceiling it has view of best and the function of self-sufficiency with sola cell.

Features of the roof of the afternoon when it is quiet and The features which takes care of vegetable from the greenhouse inside which the warm sunlight sheds light

To form the community of the roof, space of being understood and vegetable space formations of self-sufficiency.

GIVE back to YOUTH

젊은 세대에 밀려 갈곳 잃은 노인들에게 그들만의 문화공간을 제안한다

BACKGROUND

노인은 문화를 창조 계승하여 사회를 발전시키는데 공헌하여 온 사회의 어른으로서 존경받으며 안락한 노후를 누릴 권리를 가지고 있다. 하지만 정작 문화를 창조한 노인들은 사회구조, 젊은 세대의 가치관 변화에 떠밀려 문화에서 소외되어 생활의 욕구와 즐거움을 찾지 못하고 무기력해져 가고 있다. 우리나라의 인구 고령화 현상이 어디에서도 유래를 찾아 볼 수 없을 만큼 빠르게 진행됨에 따라 문화적으로 소외된 노인인구가 폭발적으로 늘어 날 것으로 예상된다. 노인복지에 대한 관심이 높아지고 있는 오늘날 경제적 지원문제뿐만 아니라 노인의 사회적 참여와 문화적 활동 또한 함께 관심을 가져야 할 것이다. 대학로 예

술가의 집 옥상에 노인들의 직접적인 참여를 이끌어 낼 목적으로 변화 가능한 다목적 공간을 제안한다.

Old generation, the society succeeded in developing the creative contribution of the whole society, respected as an adult has the right to enjoy a comfortable retirement is. However elderly who created social structure and culture. Young people alienated from the values, culture, pushed for changes in appetite and enjoyment of life is being able to find helpless. Population aging in Korea

where you can not even find the resulting fast-paced as culturally isolated; as the elderly population is expected to in crease exponentially. Exalted for elderly care issues that support today's economic and social participation of the old generation as well as cultural activities will be concerned too. Daehak-ro on the roof of the house of the artist direct participation for the older people can lead to multi-purpose space available for the purpose of the change is proposed.

CONCEPT

무기력한 삶에 직접적인 구축 작업을 통한 희열과 즐거움 자립
적 생활방식을 형성
무질서한 조합으로 인한 리듬과 변화
노인 주체 공간

Construction work directly on a helpless life of joy and self-reliant way of life by inducing the formation
Rhythm and changes due to disorderly combina
tion
Senior principal space

4가지 색으로 된 타일 104개를 섞은후 14씩 가져간다.
Four kinds of color tiles 104 aof colors, and then takes four-
teen per

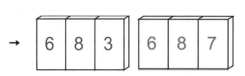

If three or more same color and same number, you can figure out
from the frame
(First person who extract all the cubes to win one game)

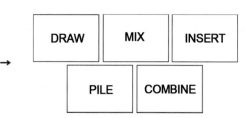

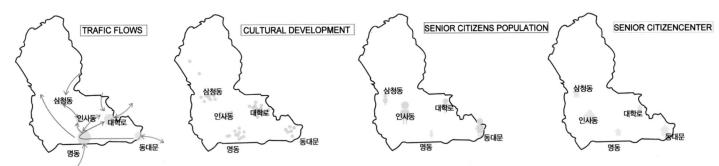

TRAFIC FLOWS · CULTURAL DEVELOPMENT · SENIOR CITIZENS POPULATION · SENIOR CITIZENCENTER

1985년,대학로라는 이름이 처음 사용됐다. 정부 주도로 문화 예술의 거리를 조성하면서 사용된 명칭은 지금까지 이어진다. 대학가 문화가 주를 이뤘던 곳으로 신촌을 비롯해 서울 곳곳에 흩어져 있던 문화단체와 극장들이 모여들기 시작했고 2004년 에는 인사동에 이어 서울에서 두 번째 문화지구로 지정돼 서울 의 문화를 대표하는 거리가 됐다.

In 1985, the first time the name was used Dae-hak-ro. The government led by the composition of culture and art as the streets now lead to the designations used. These have achieved a state where college culture, including Sinchon, Seoul, were scattered all over began to gather cultural organizations and theaters. In 2004, Daehak-ro was designated the second cultural district, as a cultural representative streets of Seoul.

ARTIST'S HOUSE

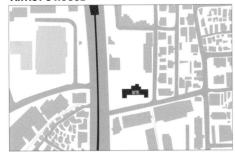

문화예술진흥을 위한 사업과 활동을 지원함으로써 모든 이가 창조의 기쁨을 공유하고 가치 있는 삶을 누리게 하는 것을 목적 으로 설립되었다. 궁극적으로 국민 모두가 문화예술이 주는 창 조적 기쁨으로부터 소외되지 않도록 하기 위해 노력한다. 하지 만 이곳 또한 대학로 여느 곳과 같이 젊은 세대들이 향유할수 있는 문화 사업에 집중하고 있다. 처음 예술가의집 설립 목적을 살려 문화예술의 기쁨으로 부터 소외된 노인 들을 위한 새로운 문화공간을 구성한다.

For the promotion of cultural projects and activi-ties by supporting the creation of a plea sure to share with everybody else and deserves to enjoy a life that was established for the purpose. Ulti-mately, all the people from the cultural arts cre-ative happiness that seeks to prevent alienation. Whenever you can see, the Daehak-ro to enjoy the culture also young people are focusing on busi-ness. The Artist House revive the happiness of art, and culture from the isolated elderly people is proposed for a new cultural space.

TARGET

THE CONTACTS

CONTACT

DAEHAKRO
↓
IHWADONG
DONGSUNG
HYEHWADONG

LABOR
YOUTH CULTURE

LONELINESS

DAEHAKRO
↓
FILIPINO WORK ERS IN THE SUNDAY MARKETPLAC

COMMUNICATION

"SENIOR CITIZEN"

NEEDS

CULTURE OF THE RLDERLY

JOBS

MAIN TARGET

SUB TARGET

FACE FORMATION

SOFT+LIGHT= SAFE

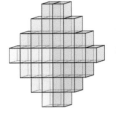

COLOR PLAN

"노인"
↓
색을 구분 하는 능력 감소
↓
진하고 화려한 색을 좋아함
↓
현란한 빛깔의 "원색"
↓
빨강, 노랑, 파랑, 초록

"SENIOR CITIZEN"
↓
REDUCED ABLITY TO DISSTINGUISH COLORS
↓
LIKES RICH AND COLORFUL
↓
OF BRILILIANT PRIMARY COLORS
↓
RED, YELLOW, BLUE,GREEN

PROGRAM SUGGESTION

A VARIETY OF PROGRAMS

MON	TUE	WED	THU	FRI	SAT	SUN
A	B	C	A	C		D

"필요에 따라 프로그램을 다양하게 구성할수 있다"
As needed, a variety of programs can be configured

DESIGN PROCESS

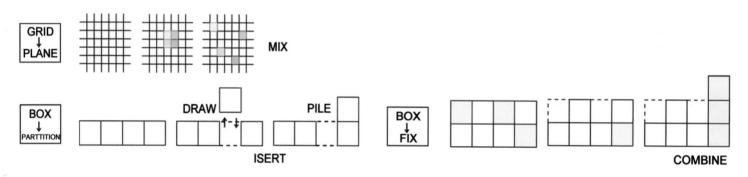

GRID → PLANE   MIX

BOX → PARTTITION   DRAW   PILE   ISERT   BOX → FIX   COMBINE

MODULE

패 ↓ MODULARIZATION

200 800 800

STEP   SEAT   PARTITION

MODUL FUNCTION

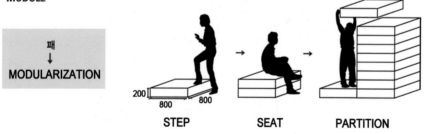

FIX
FIX
WHITE BOX

FIX   FIX
COLOR BOX

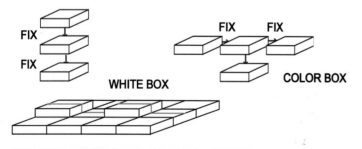

FORMATION OF THE STAIRS AND FLATS = EXERCISE EFFECTS

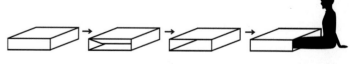

COLOR BOX TRANSFORM = EASY TO USE

316

입구 측면에 게시판 노인들의 활동 사진 등을 부착 한다.

Bulletin on the side of the entrance to the activities of the elderly can attach. photos, and more.

두뇌회전에 도움이 되는 보드게임을 통해 치매, 중풍과 같은 질병을 예방한다.

Board game to help the brain to rotate, the prevention of diseases such as stroke and dementia.

필리핀 노동자들에게 우리말과 문화를 알려주므로써 노인들 스스로 성취감을 느끼고 서로에게 말동무가 되어줄 교육 공간이다.

By providing Korean and Korea's Culture for Filipino workers, an opportunity to inform the elderly themselves to each other, feel a sense of accomplishment and that will become educational space.

공연 문화를 통해 삶의 즐거움을 찾을 수 있으며 공연 규모에 따라 공간의 크기가 달라진다.

To seek enjoyment of life through performing culture, space size canbe changed by scale of the performing.

전시공간의 사진, 그림등을 통해 노화 현상을 거부하고 젊음을 회상하게 되며 우울증을 치료하게된다.

denial of aging, reminiscent of the youth and obtained the effects of overcome depression through the Photos and pictures of exhibition space.

필리핀 노동자들에게 우리말과 문화를 알려줄 수 있는 기회를 제공함으로서, 노인들 스스로 성취감을 느끼고 서로에게 말동무가 되어줄 교육 공간이다.

By providing Korean and Korea's Culture for Filipino workers, an opportunity to inform the elderly themselves to each other, feel a sense of accomplishment and that will become educational space.

A VARIETY OF FLOOR PLANS

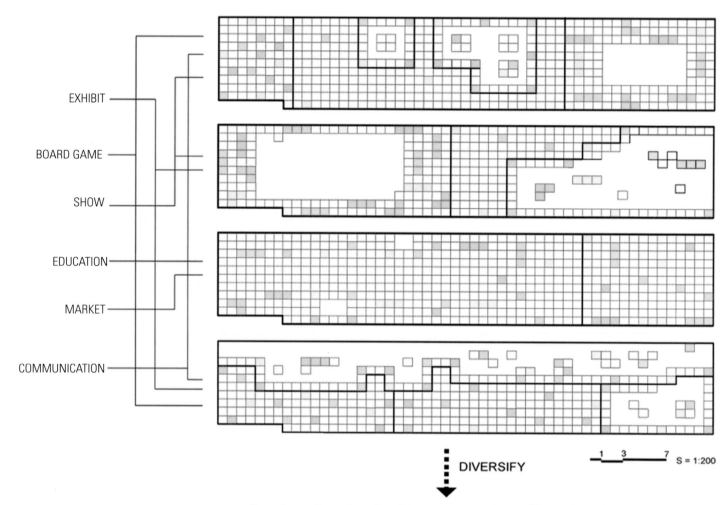

EXHIBIT

BOARD GAME

SHOW

EDUCATION

MARKET

COMMUNICATION

DIVERSIFY

1 3 7 S = 1:200

Depending on the configuration of the space program have a different size and shape.

People attending the exhibition back

ELEVATION

Likewise, plans and elevations configuration varies depending on the situation.

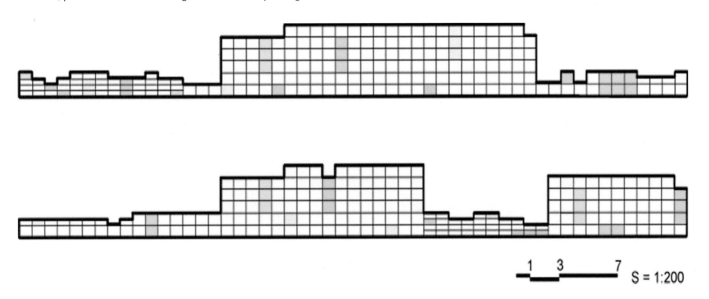

1 3 7 S = 1:200

The people watching the picture

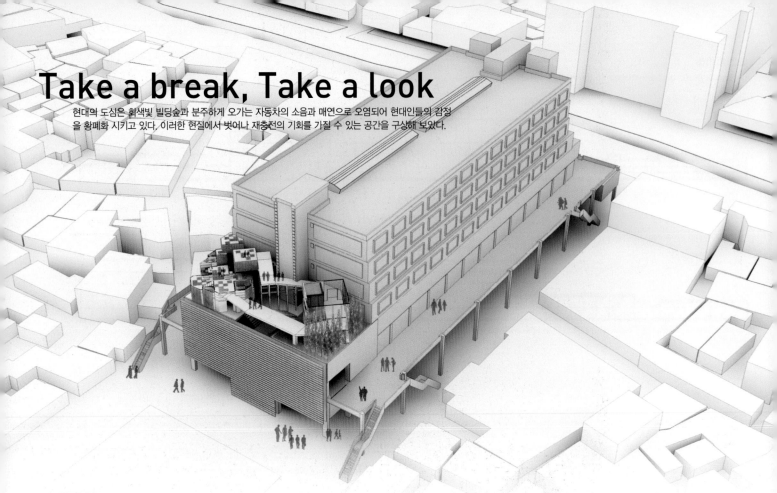

Take a break, Take a look

현대의 도심은 회색빛 빌딩숲과 분주하게 오가는 자동차의 소음과 매연으로 오염되어 현대인들의 감정
을 황폐화 시키고 있다. 이러한 현실에서 벗어나 재충전의 기회를 가질 수 있는 공간을 구성해 보았다.

BACKGROUND

현대의 도시인들은 매연, 소음 등으로 쉴새없이 괴롭힘 받고
있다. 그 중에서도 도심 한복판을 생활의 터전으로 삼아 영업
을 하는 상인들에게 있어 그들이 근무하는 상가의 환경은 가
히 최악이라 할 수 있겠다. 손님이 언제 점포에 들릴지 모르는
상인들은 잠시라도 그들의 가게를 비우는 일이 쉽지 않을 것
이다. 그리하여 이러한 상인들이 잠시 틈나는 시간을 활용하
여 쉽게 이용할 수 있는 옥상 공간을 활용해 그들의 지친 몸과
정신을 치유해 줄 수 있는 공간을 생각해 보았다.

Modern city people are suffering from exhaust
gases and loud noises from vehicles. People who
make their money for living in the middle of the
city are victims of these gases and noises. It is the
worst working environment. It would not be easy
to leave the store unattended without knowing
when the customer would come. Thus, we came
up with an idea to use a space up on the rooftop
to cure their body and spirit where they can easily
rest with their short spare time.

SITE

– 서울 종로구 종로3가 세운상가
– 상인들의 쉼터가 부족하며 기존의 옥상은 오직 흡연공간으
로서의 용도로만 활용되고 있다.
– 현재 단계적으로 상가를 철거하고 대규모의 녹지축을 조성
하려는 서울시의 계획이 실행되고 있다.

-Seoul jongro gu jongro 3rd ga, sewoon sang ga.
-Rest area is insufficient for the people who work
in the building and the space on the roof top is
only being used for smoking.
-At the moment, the Seoul city gradually put a
plan into action. The plan is to break down the
building and construct a large scale of green axis.

도심의 빌딩 숲 한복판에 위치한 세운상가는 완공된지 오랜 시간이 지났지만 유지 및 보수가 제대로 되지 않아 지역 일대와 함께 슬럼화 되어가고 있다. 이렇게 슬럼화 된 건물의 모습은 건물 뿐
아니라 주변 지역의 사람들에게까지 영향을 주게 되어 상가 주변의 모습은 어둡고 침침하다. 이러한 삭막한 공간에 활동적이고 생기가 넘치는 공간을 구성하여 분위기를 바꿀 수 있도록 계획한다.
lar maintenance and repair. So, the building is being desolated. This desolated building looks dark and gloomy and this appearance influenced surroundings
and they also became dark and gloomy. We plan to construct an active and a coruscating space to change the desolate atmosphere.

CONCEPT

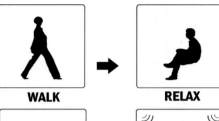

WALK → RELAX

SEE → LISTEN

안과 밖의 모호성. 흔히 우리는 안과 밖을 결정하는데 있어 논리적 사고를 통해 구분하기 보다는 시각, 청각 등의 직관적인 감각으로 판단한다. 이러한 점을 이용하여 사람들의 시선, 촉감, 후각, 청각을 이용하여 안과 밖의 공간의 경계를 허물고 자연스럽게 외부와 내부가 연결되는 공간을 구상해 보았다.안과 밖의 모호성. 흔히 우리는 안과 밖을 결정하는데 있어 논리적 사고를 통해 구분하기 보다는 시각, 청각 등의 직관적인 감각으로 판단한다. 이러한 점을 이용하여 사람들의 시선, 촉감, 후각, 청각을 이용하여 안과 밖의 공간의 경계를 허물고 자연스럽게 외부와 내부가 연결되는 공간을 구상해 보았다.

Ambiguousness of its state of inside and outside: More often than not, we rather use our senses, such as visual sense and auditory sense, than use logical thinking when we decide a space whether it is inside or outside. With this provision, we planned a space where the inside is naturally connected to the outside breaking down the wall between inside and outside.

KEYWORD

LAYER : 켜켜로[층층이/겹겹이] 놓다[쌓다]

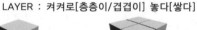

MASS : (정확한 형체가 없는) 덩어리[덩이]

DIAGONAL : 사선(대각선)으로된 매스

MOTIVE

Dynamic movements of the people in the area will lead them to engage in a lot of activity. This dynamic movement will help people to vitalize the mental tiredness caused by walking around the complicated and busy building. Also, customers will be able to see and enjoy the displayed products.

PROCESS

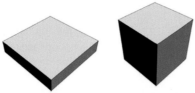

Cube is used, which is hexahedron.

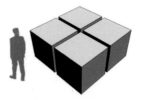

Front and side views are both utilized for arranging the cube in various direction and level.

The flow of human traffic is utilized with the spatial configuration of the rooftop and cubes that are connected in various directions.

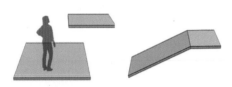

Vertically parceled space is connected using stairs with lamps.

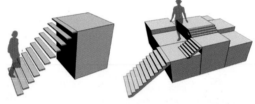

The lamps are evenly and largely distributed so that the space for human traffic is enlarged to use walking space.

전시 및 산책 공간

자연스럽게 동선을 따라 이동하면서 사이사이에 위치한 전시 부스를 관람 할 수 있다.

exhibition

The space for walk as well as exhibit walking through the naturally made path, we can enjoy the exhibitions evenly displayed along the path.

휴식공간

5층 옥상의 휴식공간 위로는 다양한 레벨의 공간을 연결하여 지루함을 덜어주고 녹지가 조성되어 있어 인공적인 공간 속에서 자연을 느낄 수 있다.

Rest Area

Above the fifth floor on the rooftop, the tedium of typical building is relieved and the nature can be felt in artificial space.

휴식공간

새소리, 물소리등과 함께 시각, 청각, 촉각 등으로 자연을 느끼며 휴식을 취할 수 있다.

Rest Area

Providing rest area for the people who work in the building with beautiful sound of bird and water, and pictures of woods on the wall projected from a slide projector so that people can rest and enjoy the nature with their ears and eyes.

전시장

세운상가의 본래 기능인 전자상가의 이미지를 이용하여 신제품의 IT기기를 전시할 수 있는 공간을 계획한다. 이 공간은 홍보의 목적으로만 이용되고 수요계층을 자연스럽게 상가로 유입시킨다.

Exhibition room

Sewoon sang ga was originally designed for digital shops. Therefore, in this room, reviving original purpose for this building, new IT products are displayed only for display, but not for the sales. This room leads people's footstep to the shops

FLOOR PLAN

옥상공간의 개방성을 확보하고 다양한 방향으로 배치된 큐브를 연결하여 동선을 형성한다.

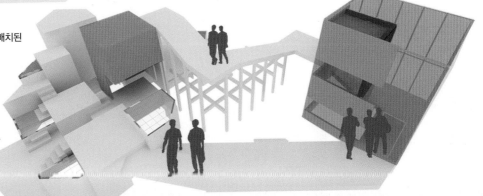

복잡하고 분주한 상가 내에서 지친 몸과 마음을 역동적으로 순환되는 이동 동선을 이용하여 시민들의 활동 심리를 유발하고 이러한 활동을 통해 상가 주민들의 정신적인 치유를 도와준다. 또한 외부에서 유입되는 시민들 역시 이러한 동선을 통해 전시된 제품을 구경하고 즐길 수 있도록 계획한다.

Dynamic movements of the people in the area will lead them to engage in a lot of activity. This dynamic movement will help people to vitalize the mental tiredness caused by walking around the complicated and busy building. Also, customers will be able to see and enjoy the displayed products.

휴식공간 위로는 다양한 레벨의 공간을 연결하여 지루함을 덜어주고 녹지가 조성되어 있어 인공적인 공간 속에서 자연을 느낄 수 있다.

Above the fifth floor on the rooftop, the tedium of typical building is relieved and the nature can be felt in artificial space.

세운상가의 상인들을 위한 휴식공간과 함께 외부로부터
유입되는 이용자를 위한 휴식 및 전시공간

The rest and exhibition are for the shop owners of
the Sewoon sang ga as well as for the customers

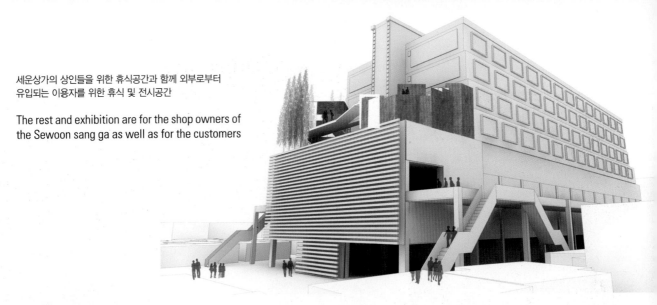

세운상가의 본래 기능인 전자상가의 이미지를 이용
하여 신제품의 IT기기를 전시할 수 있는 공간을 계
획한다. 이 공간에서는 판매는 하지 않으며 홍보의
목적으로만 이용되고 수요계층을 자연스럽게 상가
로 이동시킬 수 있도록 계획한다.

Sewoon sang ga was originally designed for
digital shops. Therefore, in this room, reviving
original purpose for this building, new IT products are displayed only for display, but not for
the sales. This room leads people's footstep to
the shops.

새소리, 물소리와 함께 시각, 청각, 촉
각으로 자연을 느끼며 휴식을 취할 수
있다.

Providing rest area for the
people who work in the building with beautiful sound of
bird and water, and pictures of
woods on the wall projected
from a slide projector so that
people can rest and enjoy the
nature with their ears and eyes.

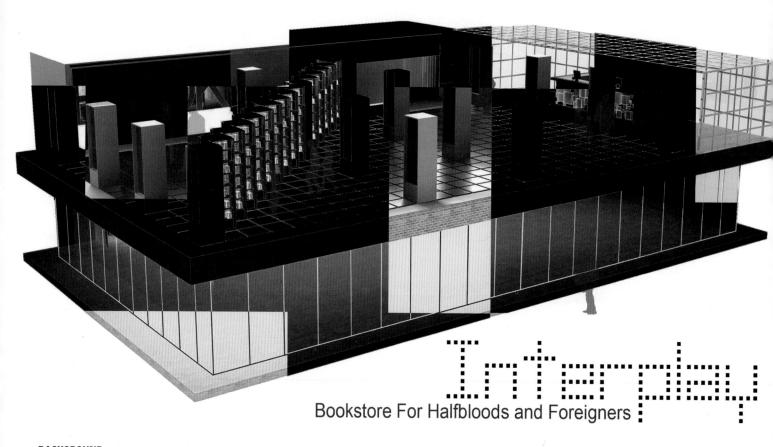

Interplay
Bookstore For Halfbloods and Foreigners

BACKGROUND

외국어와의 언어적 표현차이로 인한 이해력과 감정전달 감소를 막기위해, 각 외국인들의 자국어로 된 책을 판매의 필요성을 느낌. 또한, 독일문화원에 있는 도서관에서 소장하고 싶은 책을 발견했을 때, 바로 윗층에서 구입가능 한 서점을 선택함.

No translation is perfect because of each language`s grammar and expression. For example, since the German people living in Korea cannot fully understand Korean books due to cultural difference, the sale of books written in Germans may help them understand well.

TARGET : Halfbloods & Foreigner

어느 나라에도 온전히 속하지 못하고 교육을 제대로 받지 못하거나, 취업할 때 어려움 등 차별 대우를 받는 Halfblood. 그리고 다르게 생긴 외모로 눈초리 받고, 차별 받는 외국인들. Halfbloods와 외국인을 사회적 약자로 보고 타겟을 정함.

Practically, mixed-blood people in the world not only have hard time to get a job, but also suffer from severe social discrimintion.Therefore, we regard them as second-class citizenand target them.

SITE

– 주소 : 주한 독일 문화원
– 주변 : 근처 버스 정류장에서 바로 이태원으로 넘어 갈 수 있음.외국인들이 많은 이태원과 밀접해서, 호환성이 좋음. 그리고 관광지역으로 지정된 명동도 가까워 외국 관광객 유치도 가능함. 또한, 주변에 도서관도 많아서 애독자들의 많은 방문 예상.
– 지형 : 가장 높은 건물에서 가장 낮은 건물까지 경사가 있음. 건물을 다 오를 필요 없이 가장 높은 건물 쪽 입구에 경로가 하나 더 있어 옥상 진입이 용이함.
– 면적 : 590㎡
– 레벨 : 4m

- Site : Goethe-Institut
- Analysis : This place is located near Itaewon and Myeong-dong in which there are famous shopping malls and libraries. Foreigners will love this place because they can enjoy shopping clothes and buying books easily. Therefore, we expect them to love to visit this place.
- Topography : the slope between two buliding helps people to get close to the entrance of the buliding.
- Area : 590㎡
- Level : 4m

KEYWORD

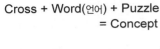

Cross + Word(언어) + Puzzle
= Concept

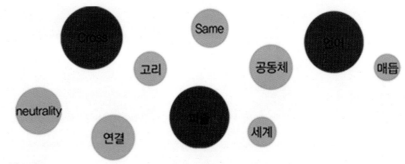

CONCEPT

: Cross Word Puzzle

글자를 넣는 칸은 흰색, 넣지 않은 칸은 검정색이다. 세로 또는 가로의 글자는 어디선가 교차한다. 따라서, 어떤 글자가 잘 풀리지 않는다해도 교차하는 다른 글자가 실마리가 되어 해답을 얻을 가능성도 있다. 크로스워드는 힌트가 문장으로 되어 있다.
[바둑판처럼 생긴 네모 속에 힌트에 의해서 글자를 채워 넣어 세로와 가로로 말이 연결되게 하는 퍼즐]

Characters put the white part, Characters do not put the black part. This is a crossword. A crossword is a word puzzle that normally takes the form of a square or rectangular grid of white and shaded squares. Therefore, you have to fill the white squares with letters, forming words orphrases, by solving clues which lead to the answers.The shaded squares are used to separate the words or phrases.

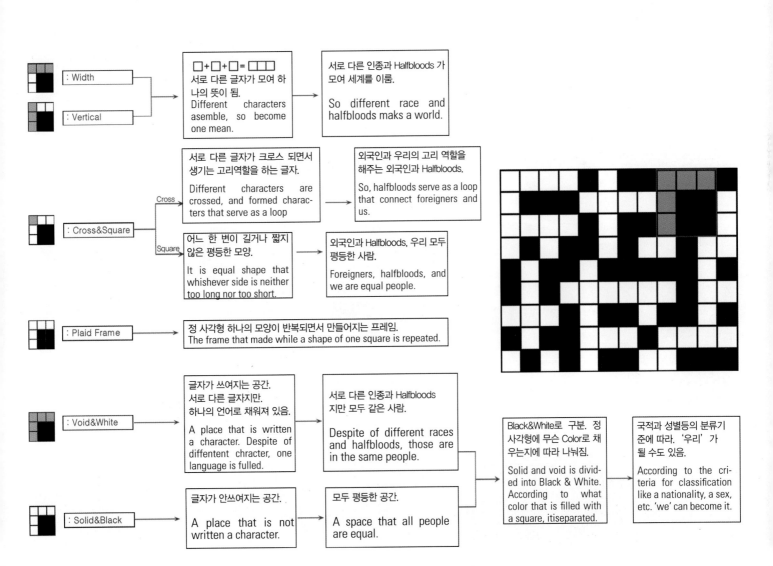

: Width

: Vertical

□+□+□=□□□
서로 다른 글자가 모여 하나의 뜻이 됨.
Different characters asemble, so become one mean.

서로 다른 인종과 Halfbloods 가 모여 세계를 이룸.
So different race and halfbloods maks a world.

: Cross&Square

Cross
서로 다른 글자가 크로스 되면서 생기는 고리역할을 하는 글자.
Different characters are crossed, and formed characters that serve as a loop

외국인과 우리의 고리 역할을 해주는 외국인과 Halfbloods.
So, halfbloods serve as a loop that connect foreigners and us.

Square
어느 한 변이 길거나 짧지 않은 평등한 모양.
It is equal shape that whishever side is neither too long nor too short.

외국인과 Halfbloods, 우리 모두 평등한 사람.
Foreigners, halfbloods, and we are equal people.

: Plaid Frame
정 사각형 하나의 모양이 반복되면서 만들어지는 프레임.
The frame that made while a shape of one square is repeated.

: Void&White
글자가 쓰여지는 공간.
서로 다른 글자지만.
하나의 언어로 채워져 있음.
A place that is written a character. Despite of diffentent character, one language is fulled.

서로 다른 인종과 Halfbloods 지만 모두 같은 사람.
Despite of different races and halfbloods, those are in the same people.

: Solid&Black
글자가 안쓰여지는 공간.
A place that is not written a character.

모두 평등한 공간.
A space that all people are equal.

Black&White로 구분. 정 사각형에 무슨 Color로 채우는지에 따라 나눠짐.
Solid and void is divided into Black & White. According to what color that is filled with a square, itiseparated.

국적과 성별등의 분류기준에 따라. '우리' 가 될 수도 있음.
According to the criteria for classification like a nationality, a sex, etc. 'we' can become it.

DESIGN PROCESS

VERTICAL
vertical upright,
sheer, perpendicular, straight
erect, plumb, on end, precipitous, vertiginous.

WIDTH
The width of
something from one is
the distance it measures edge
to the other.

Likewise, different bookshe
lves assemble.
hose make a one spacethat is
exhibition, theater, studyroom.

CROSS
Someone who
is cross is rather angry or
irritated. I'm terribly cross with

,We make a searching place to
the cross part, and fix it.
This idea of point is that bookshelf assemble one place.

SOLIDE
If you describe
someone as solid , you
mean that they are very reliable

BLACK
Something that
is black is of the darkest
there is no light at all.

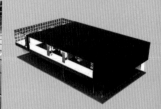

A passage that all people
pass by a stairs, a floor,
a ceiling

VOID
If you describe
a situation as a void,
you mean that it seems empty

WHITE
Something that
is white is the colour
of snow or milk. COLOUR

Despite of different races, it is
same that all people do
privacy this place. A space that is
filled with books, a space that
search books.

PLAID
Plaid is material
with a check design on it.
Plaid is also the design itself.

FRAME
The frame of
a picture or mirror
is the wood, metal, or plastic

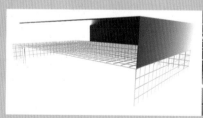

Ceiling and wall frame,
rail's shape,

Staircase's wall and
ceiling frame.

SQUARE
A square is a shape
with four sides that are all the
length and four corners that are all right angles.

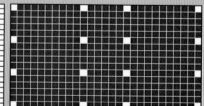

cross stripes makes a basic out-
line, so this is a basic shape of
crossword. Ceilling frame, rail,
bookshelf and a searching place
top view

DESIGN PROCESS

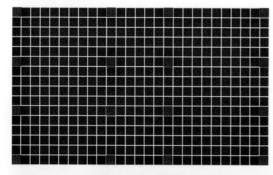

 Part of fixed　　　　 part of grid

빨간 부분은 고정 부분으로 크로스 워드 퍼즐의 흰색 부분으로 크로스 되는 부분이다. 이렇게 고정된 부분은 서점에서 찾는 도서를 검색할 수 있는 검색창으로 이용할 수 있다. 그리고 아래있는 사진들은 검색창을 제외한 움직이는 책장의 움직이는 활동 방향과 어떻게 해서 움직이는 지를 보여지는 그림과 사진들이다.

As part of the red is the part of the fixed. That part is Cross in a cross-word puzzles of the white parts. This fixed part is search box, you can search the books in a bookstore. And belowing the pictures show that how can be moved bookshelves and how does spaces are used by being moved the bookshelves.

책장의 움직임으로 인해 Black 공간이 white가 될 수도 다시 Black이 될 수도 있다.

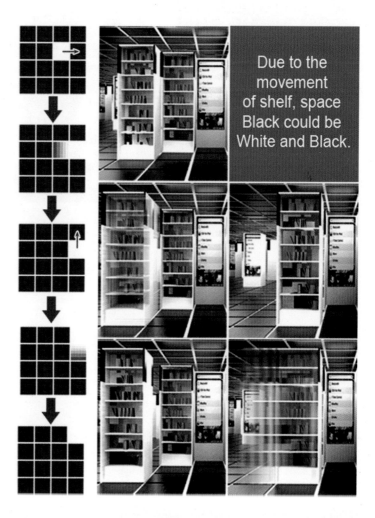

Due to the movement of shelf, space Black could be White and Black.

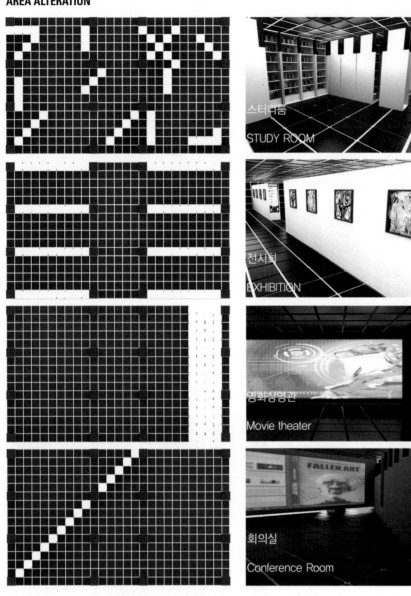

스터디룸
STUDY ROOM

전시회
EXHIBITION

영화상영관
Movie theater

회의실
Conference Room

ENTERANCE IN TO THE ROOFTOP BOOKSTORE - Bookcase stairs

INSIDE OF THE BOOKSTORE - Counter, Office, Bookshelves

Brain of City
Rocking Info

많은 정보들을 필요로 하는 도시 속 현대인들을 위한 소통의 창
exchangeable window for people needed more and deeper information

BACKGROUND

A Creation of an INTERATIVE SHARING SPACE, where the peoples crowded in the region and its neighborhood, enabling the mass COMMUNICATIVE INTERACTIONS taking place between the peoples in the territory. In open exterior of the sharing space , Peoples will be able to be accessed, or to be transmitted the informations across multiple mass display screens on the buildings surrounded.

많은 사람들이 지나가는 이 일대에 하나의 공유공간을 만들므로써 사람들의 소통을 가능하게 한다. 특히 신문사건물에 있는 전광판이라는 특징을 살려 공정한 언론으로서 사람들에게 질 높은 정보를 제공하고 상호작용이 가능하게 한다.

In interior, through the exhibitions of the materials related new media, the mass of people will be experiencing and sharing those informations as well. (Particulary, The mass display of the news paper companies surrounded neighbor, will be able to provide the people in the space the high quality information with their swift and fairness.).

외부 공간에서는 전광판으로 정보를 전달하기도 하고 습득하기도 하고, 내부 공간에는 새로운 미디어에 관련한 것을 전시하고 체험할 수 있는 공간을 만들려고 한다. (특히 신문사건물에 있는 전광판이라는 특징을 살려 공정한 언론으로서 사람들에게 질 높은 정보를 제공하고 상호작용이 가능하게 한다.)

This aims to make the sharing place possible to create mass interaction voluntarily among those peoples by the displays, acquisitions, transmissions and the experiencing of the informations and people's interests. Than the crowded solitude rampant and spread worldwide in the most of metro cities in the world.

SITE

세종로 1번지 가로로 길이 0.6km, 너비 100m
수도 서울과 한국의 정치 · 경제 · 사회 · 문화를 상징하는 중심도로 왕복 16차선으로, 도로변에는 광화문, 경복궁, 국립민속박물관이 자리하고, 도로 양옆에는 정부 주요기관과 공공기관, 대사관이 있다. 그밖에 기업과 관련한 건물들이 있다.

The most downtown crowed
Main Street No. 1 SEJONGRO
Length : 600 Meter
With : 100 Meter
Emblem of Capital City Seoul and Korea,
Symbolic main street of Politics, Economics, and Society, Dynamic Culture. Mutli-Wide street, easy to access to The Great Gate of GuangHwa, KyungBok Palace and National Fork Museum and Square and a few parks arround . As well, lots of bulidings of Governments, Institutions and Embassies, including Enter

PUBLIC INSTITUTION   COMPANY   THOROUGHFARE   CULTURAL MEDIUM

DESIGN CONCEPT

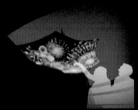

디지털 미디어의 인터넷이라는 매체를 통해 상호작용이 가능한 공간을 만드는 것이 핵심이다. 도시를 흘러 다니는 정보들을 한 데 모아 더 가까이서 접하고 감성적인 것까지 느낄 수 있는 공간을 만드려고 했다.

The core concepts is to build a SPACE where the interactions pole through the mediums of Digital media and Internet. A unique space can gives the people to feel and experience all the informations on spot , gathering informations by locally or randomly drifting over the city

MEDIA 매체   +   HUMAN 사람   >>   SPACE 공간

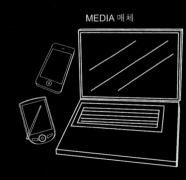

Nowadays, all direction networking available and popular by the telecommunications device and equipments.

The more info needed, and it lead communicates farther and more peoples.

Agenda to provide A window where exchangable for the people and as A space where the information easy accessable as well or communicators between the people.

MOTIVE

뇌 중 최고의 중추이고 고등사고의 영역은 대뇌피질에 해당된다. 정보를 인지하고 받아들이고 조합하고 대응하여 판단하는 역할을 모두 한다. 이 대뇌피질에서 많은 정보의 흐름이 일어나는데 많은 뉴런들의 결합이 일어나고 정보를 저장할 면적이 많아지면서 한정된 두개골 안에서 뇌는 표면적을 넓히기 위해 불규칙적으로 구불구불한 모양을 띄는데 그것이 뇌의 형태가 되는 것이다.

MOTIVE

The center of the thoughts is the Human brain ,and its domain is the CEREBRAL CORTEX(surface layer of the brain) . where charges all the Recognitions, Acquirements and combinations ,correspondence and judgements done. In This part of the brain ,As more increase massive volume of the informations flows, the more neurons should be combinated and it needs more spaces to store the informations on limited surface of Serebral Cortes on inside of the skull. That is the reason why The human brain shaped irregular curved and winkles on the surface through the evolution of the million years, to give more space , now the human brain's out look gives.

DESIGN PROCESS

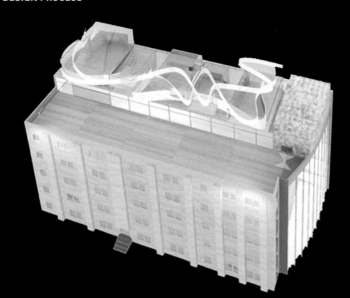

이 프로젝트는 이 공간으로 하여금 사람들이 정보를 얻기도 하고 주기도하는 상호교환을 할 수 있게 하는 곳이다. 그 인터렉션을 가능하게 하는 부분이 우리 몸에서 뇌이기에 전체적 형상을 뇌 피질에서 따와 유기적 곡선의 형태가 나왔다.
사람들이 이 공간에서 내부이자 외부를 함께 느낄 수 있고 하고 더 많은 공간을 사용하기 위해 높이와 위치를 자유롭게 배치한 박스로 공간을 구성했다.

The framework of this project is to make the SPACE to acquire and to deliver the informations among the people to exchange mutual. The general form of the space has been taken symbolic from the shape of our brain, because all the decisions and interactions are undertaken by our brain.
Cubics BOXES arranged and lined with various combinations. (To make more space and approach in and out side as well, Orgarnized agrrangements by freely height and location= by vertically and horizontally)

Magnify the cerebrum

Forming lines by the Magnified curve and winkle

Formed line drawed to intimately simplified lines

Simplified lines turns cubic effect with Mass display (Electric display board)

Sequential same boxes line up

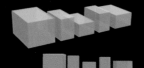

Transformations of the box sizes (Different sizes)

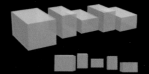

Vertical pile up of the boxes

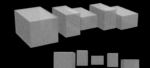

Horizontal disposition of the boxes

SPACE PROGRAMMING

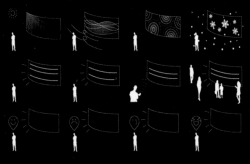

공간은 역할에 따라 2가지로 나뉜다. 하나는 전광판으로 그 일대의 사람들이 모두 볼 수 있는 화면이다. 2차원적인 화면에 여러 가지 프로그램들을 설정해 날씨, 뉴스, 시사, 경제, 다큐멘터리 등을 제공하며, 정보 제공뿐만 아니라 사람들의 의견을 받아들일 수 프로그램도 제공하여 하나의 상호교환이 가능한 3차원적 공간으로 만드려고 한다.

Divided 2 spaces by its roles. 1st one is the electric display, which the peoples neighbour in the space can see the informatins on spot from the

Displays on , or the top of the number of the buildings arround. By the various program settings, not only it will provide the whethers, news, issues and economics, documentaries on 2 dimensional monitors 1st. But also, it will be possible to do interactive exchange on demand of each individuals in the space. The space will be escalated to 3 Dimensional stage development, By the on demand interactive devices installed.

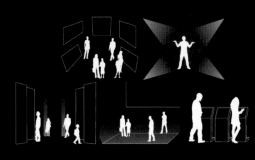

또 다른 공간은 전시공간이자 안내소와 같은 역할을 하는 곳으로 미디어에 관련한 전시 프로그램들을 설정해 제공한다. 직접 정보들을 모으고 정리해서 습득할 수 있다. 표현하고자하는 미디어 관련 내용 들을 기술적으로 풀어내 사람들의 감성을 자극하고, 이런 내용들이 더 가깝게 느껴질 수 있도록 하려한다.

The other 2nd role is exhibition pavillion stated box, or boxes of above column DESIGN PROCESS , also can be information desk for the people visited at the same time. This will be providing the Exhibition programs related media with vari-

ous contents where the individuals can collect, combine the informations on their interests for their acquirements and appreciations even can be shared among individuals on spots by themselves, also by on demand interactive devices provided , or by their devices as well. This is the ultimate agenda to make a space which can stimulate the people's sensitivities and sensibilities through the intended media contents technically set among the mass free in the SPACE for their own INTER-ACTION and to give close INTIMACY FEELING.

PLAN

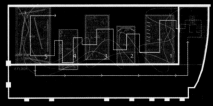
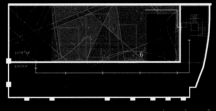

1. GLOBAL TOPIC   3. MEDIA   5. VEDIO CLIP DISPLAY
2. EXPRIENCE   4. SEARCH   6. INFORMATION CENTER

ELEVATION

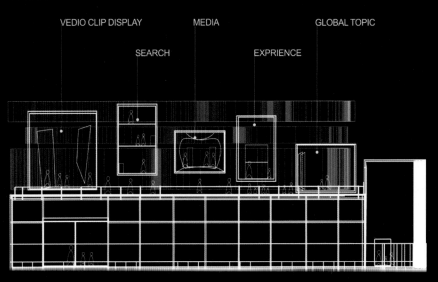

VEDIO CLIP DISPLAY   MEDIA   GLOBAL TOPIC

SEARCH   EXPRIENCE

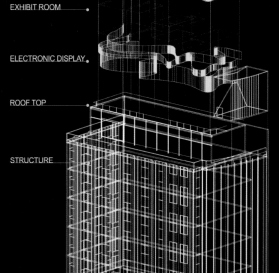

ELETRONIC DISPLAY

EXHIBIT ROOM

INFORMATION SPACE

OVERLAP

EXHIBIT ROOM

ELECTRONIC DISPLAY

ROOF TOP

STRUCTURE

ROOF TOP
01 VEDIO CLIP DISPLAY
02 SEARCH
03 MEDIA
04 EXPRIENCE
05 GLOBAL TIPIC

06 FLOOR

GLOBAL TOPIC
A space where the deeper search available on the particular issue and topics matters. nowadays.

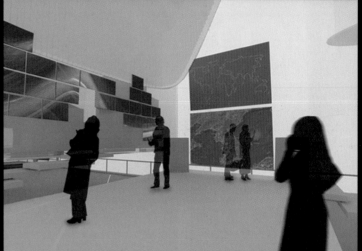

EXPRIENCE
Imaginable space where we can explore the new digital devices & equipments.

In open exterior of the sharing space. (Particulary, The mass display of the news paper companies surrounded neighbor, will be able to provide the people in the space the high quality information with their swift and fairness.)

MEDIA
The place where combined with written daily news on display gather opinions and give participates opinions as well.

SEARCH
Place where individual interests search available.

IMFORMATION DESK
Systematic information space achieve people can find their interests, funs, and knowledges available without any guide, or guidance.

VIDEO CLIPS DISPLAY
The place multiple interest video clips shown, and can give a communication center between the participants.

The Sukkah for Examinees

고시준비생을 위한 튜터 & 스터디 용도의 커뮤니티 베이스
Community base as a tutor and study tool for students who prepare for national examinations

SUKKAH_유대교의 Sukkoth 축제 때 식당 등으로 쓰이는 임시 초막(草幕)
A sukkah is a temporary hut constructed for use during the week

BACKGROUND

고시촌의 옥상, 기존 고시촌의 옥상 공간은 빨래를 건조하거나 에어컨 실외기 보관. 그 외에 담배를 피우는 장소로 사용되고 있었다. 본 프로젝트의 위치는 신림9동 255번지 241호 일대이다. 신림동 고시촌 중에서도 가장 고지대에 위치하고 있어 문화시설, 편의시설, 학원시설 등 근린시설과의 접근이 좋지 않으며 시설도 많이 낙후되어 있다. 고밀도 지역인 만큼 넓지 않은 옥상공간이 가지는 한계가 있다. 고시촌의 옥상, 기존 고시촌의 옥상 공간은 빨래를 건조하거나 에어컨 실외기 보관. 그 외에 담배를 피우는 장소로 사용되고 있었다.

BACKGROUND

Rooftops of many buildings where there are small accommodations for students who are preparing for national examinations, also called Gosichon, are currently used as a place for drying laundry, storing air-conditioner parts outside and smoking. This project has been conducted at 241 unit, 255 Sillim9-Dong 255. The area is located higher than any other Gosichons in Sillim-Dong, and lacks quality neighborhood facilities such as cultural facilities, amenities and academies. The area also lags behind the national standard in terms of the above. This area has high density, which means buildings there do not have spacious rooftops, and people in them can be seen easily from outside, which can lead to private infringement. However, there are clear differences between hillside Gosichons and low-lying ones. Hillside Gosichons have relatively smaller floating populations and lower capabilities of acquiring related information than low-lying ones. Thus, this project proposes community space as a place for tutoring and studying for Gosi students in unique situation of preparing for national examinations, who are just coming and going from their accomodations and academies everyday without enjoying any cultural lives.

SITE ANALYSIS

SELECTED AREA

EDUCATION SPACE

LIVING SPACE

COMMERCIAL SPACE

APPROACH

건물의 주 용도가 상가/학원(독서실)/주거용으로 뚜렷하게 나눠져 밀집되어있는 것이 특징이다.
The main purpose of building a commercial / Study room / residential is characterized by densely clearly divided

PROJECT DESCRIPTION

위치 : 서울특별시 관악구 대학동 255–241 일대
1975년 서울대가 관악캠퍼스로 이주 해 온 뒤로 고시생이 몰려들어 하숙촌을 형성했고 현재 약 3만명의 고시생이 이곳에 머무르며 국가고시를 준비하고 있다. 5층 미만의 고밀도 건물들이 빼곡히 들어서 있으며 산비탈이 많아 저지대와 고지대의 레벨 차이가 심하다.

Location : Gwanak-gu, Seoul 255-241 daehak-dong Seoul National University campus in 1975, migrated to the wind came back gosisaeng two swarms have formed village housing about 30,000 people currently staying in the gosisaeng

here are preparing for state examinations. Packed density of less than 5 floors of buildings heard a lot of slopes and lowlands and highlands of the level difference is keen.

DESIGN CONCEPT

"뿌리"

[명사] 식물의 밑동으로서 보통 땅속에 묻히거나 다른 물체에 박혀 수분과 양분을 빨아올리고 줄기를 지탱하는 작용을 하는 기관.

모든 것에는 기본이 있듯이 나무의 기초는 뿌리이다. 가는 잔뿌리는 원줄기 쪽으로 갈수록 점점 두꺼워 지며 여러 가지 갈래들이 하나의 군집을 이루다가 이윽고 곁줄기를 따라 다시 수많은 가지로 나눠진다. 앞으로 국가고시에 합격하여 사회의 각 분야에 진출할 고시준비생을 나무의 성장 과정 중 기초에 해당하는 뿌리에 대입시켰다. 따라서 국가고시를 준비하는 고시준비생을 대상으로 하나의 네트워크 망을 구축하고 튜터링과 스터디라는 목적을 부여하여 자치적인 프로그램을 만들고자 한다.

"ROOTS"

[noun] butt of the plant as a normal soil moisture and nutrients muthigeona other objects embedded in the sucking action to raise the institutions that sustain the stem.

The base of a tree is its roots as everything has its base. Its rootlets become increasing thicker as it gradually reaches its main roots, and, from there, the roots is again divided into numerous other rootlets. This project assigned the roots of the tree, which is the base in the process of the tree's growth, to the students who are preparing for national examinations, because they are going to advance into various social sectors when passing their national examinations. Thus, this project is to establish a network of the students, and create autonomous programs for their tutoring and studying.

DESIGN KEYWORD

MESH NETWORK

ORGANIC

BIFURCATION

MANIFOLD

CONCEPT PROCESS

학원에 접근하기 용이하지 않은 지역 조건과 경제적으로 넉넉하지 않은 상황을 고려하여 기존의 옥상을 이용하여 자치프로그램을 개설, 용도를 부여하고 인근 주민간의 튜터링과 스터디를 통해 비싼 수강료를 아낄 수 있도록 하는 것이다. 튜터를 해줌으로써 교학상장(敎學相長) 할 수 있고 소소한 수입을 얻을 수 있으며 튜터를 받는 입장으로서도 비교적 낮은 가격에 학습기회를 얻을 수 있어 상호 유리한 프로그램이 될 수 있다.

나무의 뿌리
Root of a tree

뿌리의 단순화
HUB와 FEEDER로 구분
Simplifying the tree,
dividing into Hub and Feeder

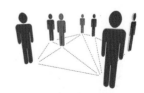

거주민간 네트워크 구축을 통해
자치적인 프로그램 시행
Administering autonomous programs by establishing
networks of residents.

DESIGN PROCESS

The lower part of cut tree

Simplifying the lower part of the tree

Binding the upper parts together

Pulling the knot inside

Divided roots

Simplified cross-section

Presenting structure by overlapped cross-sections

A role of communications and passage

Cross-section of a tree's cross-node

Simpllifying node's cross-section

Longitudinal-section of node as seen at the top

Creating a space from the shape

Simplifying the tree

Establishing as a mass

Overlapping the model

The role of message board as a post

강의실 외에 휴식공간으로도 활용 가능하다.
It can be used as a rest place as well, except as a classroom.

정보를 전달 할 수 있는 공간으로서 튜터링에 대한 정보를 많은 사람들에게 제공할 것이다.
It will provide many people with the information about tutoring, as a place in which such knowledge can be delivered.

야외 계단강의실로서 보다 많은 인원이 함께 수강할 수 있다.
As an outdoors lecture place in a terraced form, more students can take the classes in here.

브릿지는 옥상과 옥상을 연결하는 통로 역할을 하고 있다.
The bridge serves as a hallway between the roofs.

ANNUAL PLAN

	1	2	3	4	5	6	7	8	9	10	11	12(월)

행정고시 (행정직)
Civil Service Examination
　　　　(기술직)
--- 1차 ----------- 2차 ------------- 3차
--- 1차 ----------- 2차 ------------- 3차

사법고시
Bar exam
--- 1차 ----------- 2차 ------------- 3차

외무고시
Foreign Service Examination
--- 1차 ---- 2차 ---- 3차

1차 필기	2차 논술	3차 면접
Written test	Essay test	Interview

헌　법　constitution
민　법　droit civil
형　법　criminal law
행정법　administrative
상　법　commercial
논　술　essay test
면　접　interview
국제법　international law
노동법　labor law
조세법　tax law
사회법　social law
소송법　prozessrecht

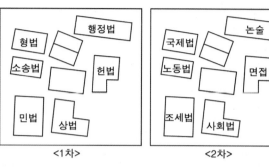

<1차>　　　　　　<2차>

연간계획 별로 시험과목을 운영하고 상관관계에 있는 법과목을 집합하여 효율적으로 운용할 수 있다.

We can efficiently operate the programs by running courses for each subject according to yearly plan above and gathering law subjects related to the students.

SPACE PROBLEM

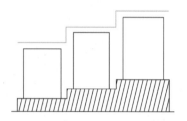

Extreme slope and different building levels

High density area and possibility of private infringement

A space where you can acquire your needed information, measures to blind others' sights and facilities to take a rest and study.
Information sharing using leaflets, sufficient blinding effects, Hub for relaxing and lecture rooms using bridge.

REQUIERMENT

Information Area　　　Privacy Window　　　Rest Area　　　Study Area

SOLUTION

전단을 이용한 정보 공유　　　적당한 시선 차단 효과　　　휴식을 할 수 있는 HUB　　　BRIDGE를 이용한 강의실　　　소규모 미팅룸

삭막한 개미마을에 꽃이 피다.

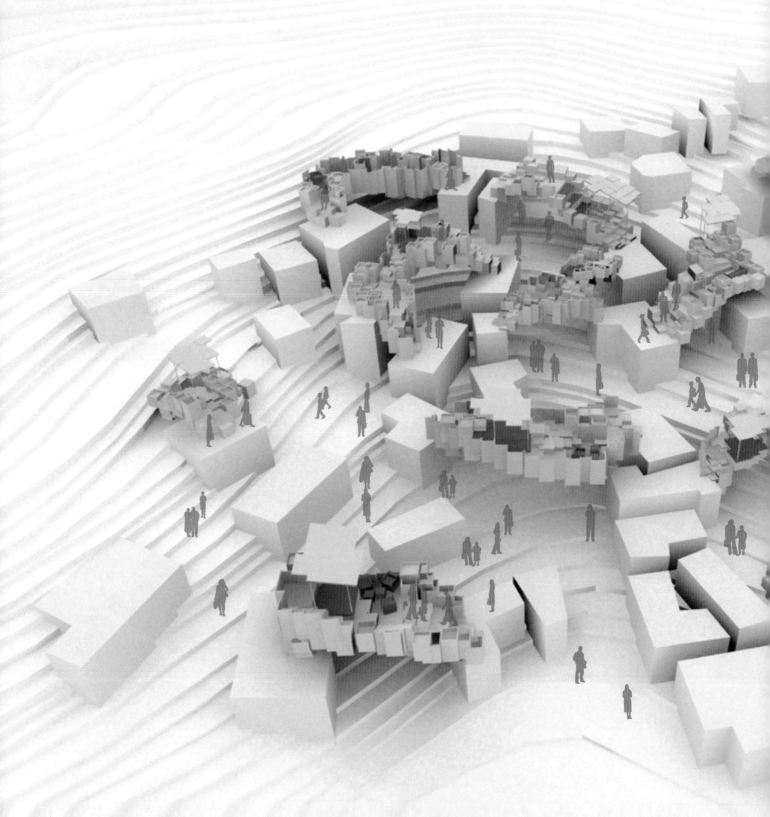

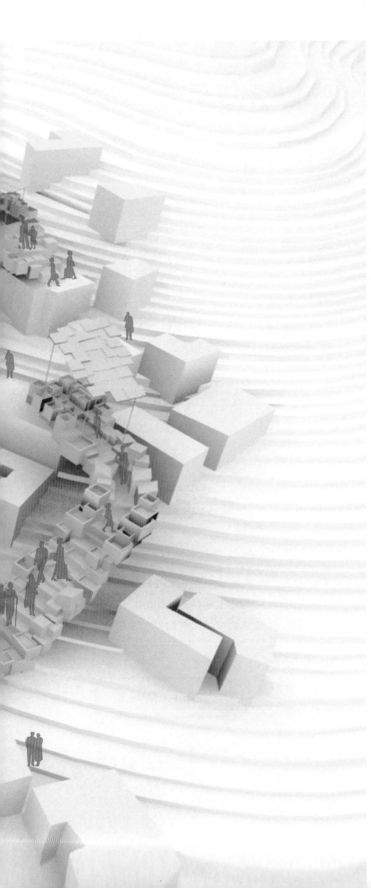

"Blooming in GAMIMAUL"

BACKGROUND

6.25 전쟁 이후 갈 곳이 마땅치 않은 가난한 사람들이 들어와 임시 거처로 천막을 두르고 살아 당시에는 '인디언 촌' 이라고도 불렸다. 1093년 주민들이 열심히 생활하는 모습이 개미를 닮았다고 해서 '개미마을' 이라는 정식명이 생겼다. 2009년 9월 5곳의 대학에서 미술을 전공하는 128명이 벽화를 그리기 시작해 지금은 문화 특구 보존 마을이 되었다.

BACKGROUND

After the Korean War, the place where all the poor people gathered and living under the tents was called 'Indian Town'. In 1093, this town is formally called 'GEMIMAUL' because the people in the village live their lives like an ant. September 5th, 2009, 128 art students began to paint a mural and now this GEMI-MAUL is specially designated as a preservation village.

SITE

주소 : 홍제 3동 9-81 개미마을
면적 : 1만 5000천평
가구수 : 210가구
주민인원 : 420명
주민평균연령 : 60~70세

SITE

Address : 9-81 GEMIMAUL, Hongje-3dong
Extent : 15,000py (12ac)
Households : 210
Population : 420
Average age : 60~70

개미마을 문제점

· 옥상이 없다.
· 빨래를 널 곤간이 없어 대문 옆이나 난간에 넌다.
· 텃밭을 좁은 길, 녹지부분에서 사용하고 있다.
· 아슬아슬하게 장독대나 물건들이 난간 위에 놓여있다.
· 주민들 서로간의 소통이 없다.

GEMIMAUL PROBLEM

· No rooftop
· No place for hanging out laundry
· They have been using vegetable garden at pathway and green section.
· Crocks and stuffs are on the railing narrowly
· No communication among neighborhood
· Wanting their rooftops to be suitable place for grow ing and exercising

KEYWORD

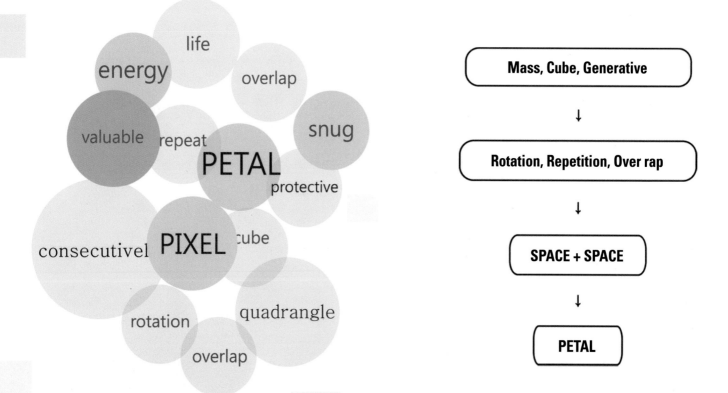

life
energy
overlap
valuable
repeat
PETAL
snug
protective
consecutivel
PIXEL
cube
rotation
quadrangle
overlap

Mass, Cube, Generative
↓
Rotation, Repetition, Over rap
↓
SPACE + SPACE
↓
PETAL

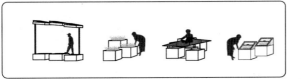

개미마을의 기존의 집

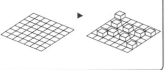

DESIGN CONCEPT

자급자족의 기능과 의미를 낙후된 개미 마을 옥상에 부각시켜 디자인 하고자 한다. 개미 마을은 옥상이 없어 좁은 집에서 모든 일들을 해결하는데 기능적인 옥상을 만들어 줌으로 써 좁은 공간에서 이루어 졌던 것들을 옥상에 서 해결한다.

DESIGN CONCEPT

We want to design that self-sufficint funtioning and meaning to lagged gemi maul's roof. GEMI-MAUL people work everything at their small house. By making the funtional roof,they can re-solve the problem of narrow space.

MOTIVE

"GENERATIVE PIXEL"

디지털 이미지를 이루는 원소.

모니터 등에 나타난 디지털 이미지의 경우 수많은 타일의 모자이크 그림과 같은 사각형 픽셀로 이루어져 있다. 픽셀이 하나하나 생성되면서 그 모양이 변화하고 커짐으로써 구조체를 이룬다.

"PETAL"

꽃에는 꽃잎과 수술이 있는데 꽃잎은 꽃부리를 이루고 있는 하나하나의 조각이고, 꽃 중에서 가장 아름다운 부분이며, 수술을 싸서 보호한다. 완성된 mass를 제시함으로써 구조체를 이루게 되어 꽃잎이 만들어진다. 그러므로 꽃잎의 구조체가 모여 한송이의 꽃이 됨으로 삭막한 개미마을에 한 송이의 꽃이 피어난다.

MOTIVE

'GANERATIVE PIXEL'

Make digital image up element.

The digital image shown up to a monitor is made up of tetragonal pixels just likea number of tile s mo-saic picture.By creating the pixels one by one, they consist of structure by enlarging and changing itself.

"PETAL"

Flowers have petal and petalody.Petal is a piece to make the corolla of a flower.

Also the most beautiful part of a flower and it protects petalody. As completed Mass introduce, the petal is making. Therefore gathering flower's structure become a flower and blooming a flower at gaunt gemimaul.

PROCESS

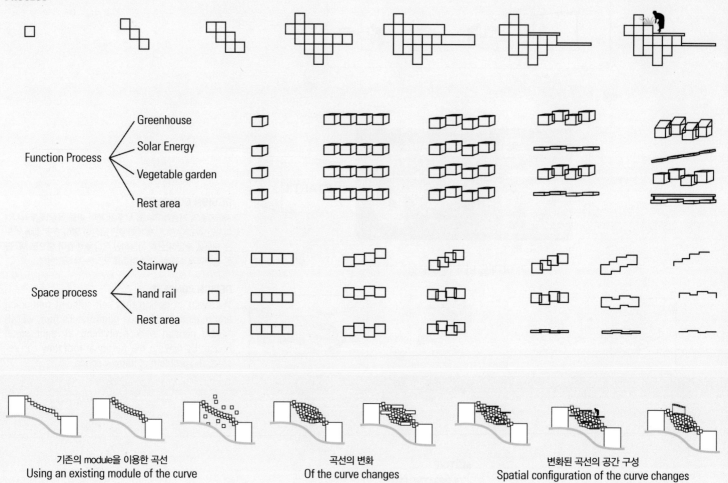

Function Process ⟨ Greenhouse / Solar Energy / Vegetable garden / Rest area

Space process ⟨ Stairway / hand rail / Rest area

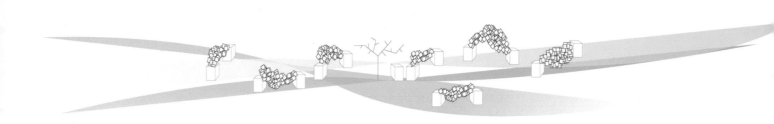

기존의 module을 이용한 곡선
Using an existing module of the curve

곡선의 변화
Of the curve changes

변화된 곡선의 공간 구성
Spatial configuration of the curve changes

DESIGN PROCESS

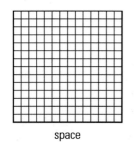

space

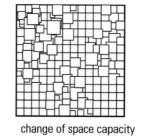

rotation of space

change of space capacity

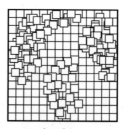

overlapping space

structure space

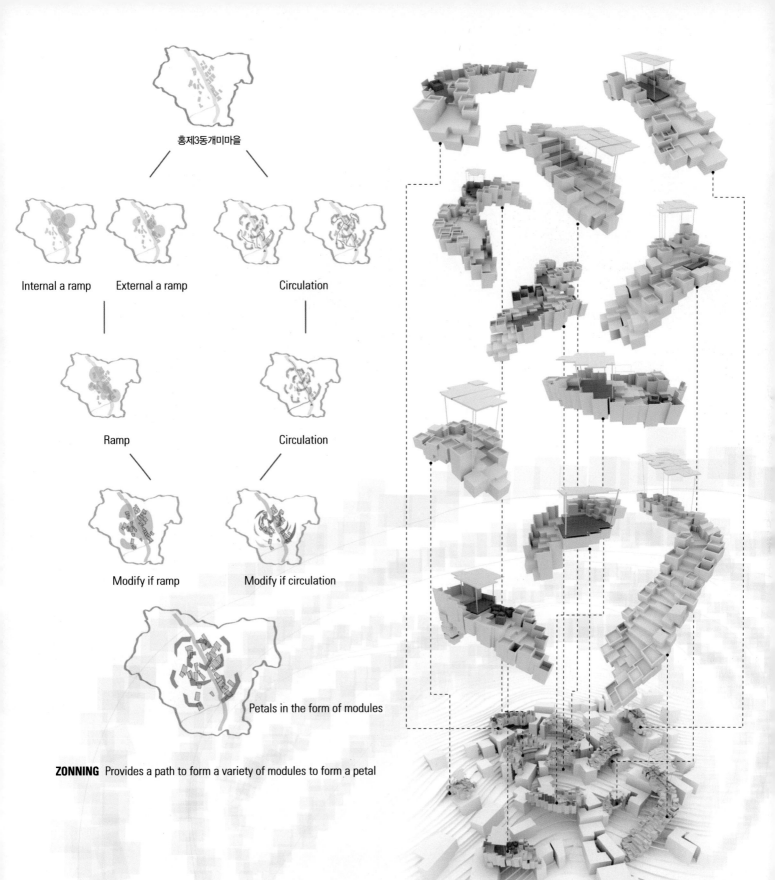

홍제3동개미마을

Internal a ramp   External a ramp   Circulation

Ramp   Circulation

Modify if ramp   Modify if circulation

Petals in the form of modules

ZONNING Provides a path to form a variety of modules to form a petal

EXPLODED VIEW Steep Ant village, people of different paths providing

349

빨래 건조대
바람이 잘 통하는 곳에 빨래대를 설치하여 옥상공간을 활용했다.

DRYING RACK
Where the wind by installing a laundry units took advantage of the roof space.

PROGRAM

자급자족 – 텃밭, 온실
· MASS에 공간이 생기면서 자급자족 할 수 있는 텃밭과 온실 또는 화분, 장독대 보관하는 공간
· 텃밭을 이루면서 필요한 물자를 얻을 수 있고, 주민들과 서로 공유 하며 상부상조 할 수 있다.

Self-sufficiency - vegetable garden, glasshouse
· By generating space on Mass,it is going to use a glasshouse, flowerpot and keep a storage well.
· Through the vegetable garden, food is provided and with sharing it people can become more friendly.

에너지 – 태양열(태양열판)
· 칸막이 공간에 태양열 판을 만들어 에너지를 얻을 수 있는 공간
· 태양열을 이용해 온수, 난방, 냉방에 이용할 수 있으며, 무한할 뿐만 아니라 깨끗하고 공해가 발생하지 않는 청정에너지이다.

Energy - solar heat
· To make solar colletor on partition space gets some energy.
· By using solar heat, people get to use unlimited hot water, heat. Besides, it's clean energy.

쉼터 – 대청마루
· 칸막이에 공간이 생기면서 개미마을 주민들이 쉴 수 있는 공간
· 대청마루를 놓으므로 사방에서 바람이 잘 통하며, 몹시 더울 때 앉아서 바람을 쐬며 음식을 먹거나 잠을 자기도 하고, 가족이나 주민들과 모여 이야기를 나누기도 한다.

Rest area - DAECHUNG MARU
· As the space generated by partiton wall,it takes a rest for GEMIMAUL's residents.
· Families and neighbours get to rest on DAECHUNG MARU, this rest area is very convenience, especially in summer, it is a great place to gather around and take a rest.

Hot water and heat is provided by solar heat and DAECHUNG MARU is rest area for the residents.

Rest area on the rooftop

Drying rack

Wood floor and solar heat

태양열을 이용해 온수, 난방, 냉방에 이용할 수 있으며, 바로 밑에는 대청마루가 있어 주민들의 쉼터이자 자급자족 공간이다. 태양열을 이용해 온수, 난방, 냉방에 이용할 수 있으며, 주변의 텃밭을 통해 신선한 야채를 필요한 만큼 얻을 수 있다. 텃밭을 이루면서 필요한 물자를 얻을 수 있고, 주민들과 서로 공유하며 상부상조 가능하며 온실을 통해 4계절 동안 계속 필요한 채소를 얻을 수 있다. 대청마루 옆에 놓인 쉼터에 앉아서 바람을 쐬며 음식을 먹거나 가족이 모여 이야기를 나누기도 한다. 경사가 가파른 개미마을의 계단 곳곳에 데크로 만든 쉼터가 자리잡고 있다. 바람이 잘 통하는 곳에 건조대를 설치했고 옆에 대청마루가 있어 빨래를 말리며 쉴 수 있게 했다. 텃밭 주변에 쉼터를 둠으로써 자연과 더 가까워 질 수 있으며 채소를 가꾸기 더 편하다.

Hot water and heat is provided by solar heat and DAECHUNG MARU is rest area for the residents. People get fresh vegetables through their own garden. By owning vegetable garden and greenhouse, residents are self-sufficient for food all along through the year. They can be connected by sharing food which is very typical in Korea. There is a rest area next to the DAECHUNG MARU where people can have a nap and eat food. There are also several rest areas on the steep slope steps of Ant Village.
Drying rack is placed at an airy spot and V is next to it so residents can take rest while drying the laundry. By placing rest areas around vegetable garden, residents can get more close to the environment and easy to grow greens.

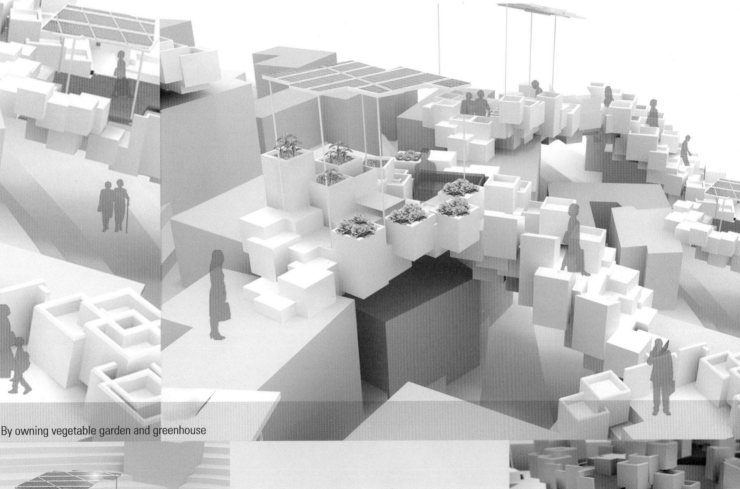

By owning vegetable garden and greenhouse

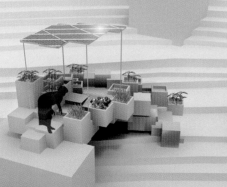

Greenhouse

Wood floor and solar heat

Vegetable garden

Boundary of public and privacy

옥탑방의 홀로서기 – 공적영역과 사적영역의 울타리 속에서.....

SITE

서울특별시 구로구 구로5동
- Gurogu Gruo 5 dong, Seoul

BACKGROUND

옥상위의 집들이 매우 규칙적으로 배열되어 집사이의 간격이 좁게 구성되어 있다. 때문에 각각의 집들 간의 시선이 차단이 어렵고 프라이버시 침해로 인한 문제가 생기게 된다. 햇빛에 그대로 노출된 집의 특성 때문에 여름 더위문제가 심각하다. 옥탑방의 좁은 집사이 문제 프라이버시 침해를 해결하기 위해 각각의 집 들 사이의 시선들을 분석하고 이를 통해 옥탑방의 근본적인 문제를 해결하고 그 외 옥탑방에 자급자족을 하는 사람들을 위한 공간을 구성하려 한다.

Houses on the rooftop was very regularly arranged, so distance in between each houses narrow. Therefore it is too difficult to block each houses' glance and these invasions of privacy causes many problem. The summer heat problem is serious because attribute of house that is directly exposed to sunlight.To solve the rooftop house of narrowing distance problem and the invasion of privacy, analyze the each houses' glance through this analyzing, solve the fundamental problem of the rooftop house and make place for self-sufficing people in rooftop house.

PROBLEM

옥탑방들이 빽빽이 밀집되어 집들 간의 프라이버시가 보호되지 못하는 단점이 있는 곳

Place that cannot protect each home's privacy because the space was in a compact mass with rooftop house.

SOLUTION TO THE PROBLEM

집들끼리의 사생활 침해로 인한 피해를 겪고 있으면서도 이사를 가지 못하는 현실의 비극적 상황에 처해있는 구로동 옥탑방 사람들을 위해 실용적 버티컬을 이용해 시선을 차단하여 풍요로운 삶을 계획한다. 높이가 일정하고 빽빽하게 밀집되어 있는 옥탑방의 특징으로 인해 사생활이 침해되기 쉬웠던 문제를 버티컬의 개념을 도입하여 이를 해결하고 자급자족적인 생활을 할 수 있도록 다양한 기능을 구성한다.

Plan the affluent life through blocking glances by using practical vertical blind for the people that realistically face tragic circumstances that cannot move into a new house while they are suffering for invasions of privacy. Using vertical blind to solve problem that is ease to infringe privacy life and is caused by attribute that height is constant and is densely arranged. Beside vertical blind affords to lead self-sufficing life for people who live in rooftop house.

SITE ANALYSIS (Between public and privacy)

옥탑방의 좁은 집사이 문제 프라이버시 침해를 해결하기 위해각각의 집 들 사이의 시선들을 분석하고 이를통해 옥탑방의 근본적인 문제를 해결하고 그 외 옥탑방에 자급자족을 하는 사람들을 위한 공간을 구성하려 한다.
To solve the rooftop house of narrowing distance problem and the invasion of privacy, analyze the each houses' glance through this analyzing, solve the fundamental problem of the rooftop house and make place for self-sufficing people in rooftop house.

CONCEPT

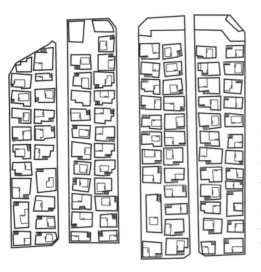

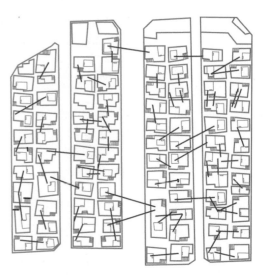

옥탑방은 서로 멀리 있을수록 프라이버시가 보호된다. 이를 옥탑방들이 서로 상호 작용을 하여 가까이 있음에도 불구하고 프라이버시가 지켜 지도록 한다.

The more each rooftop house keep away the more it is ease to protect privacy. Although rooftop house close each other to interaction, focus on protecting privacy using rooftop houses.

MOTIVE

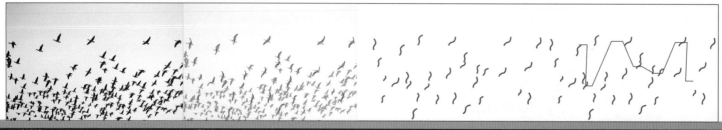

밀집성	방향성	기능성	율동성
빈틈없이 빽빽하게 모임성	방향이 나타내는 특성 방향에 따라 제약되는 특성	기능이 가지는 역활과 작용 효과의 정도	일정한 규칙에 따라 일정한시기에 변화하여 움직이는 성질

Boid

새떼처럼 비슷한 집단처럼 보이도록 1987년 SIGGRAPH에 제출된 논문에서 처음 등장한 기법이다. 모든 Boid들은 하나의 무리로서 움직이고, 장애물과 적들을 피하며 실제로는 단순한 규칙들의 상호작용들로 이루어졌을지도 모른다는 가정을 갖게 한다.

This technique first appeared at the paper submitted to SIGGRAPH in 1987 to look like similar group as birds. It has assumption that every Boids move as one group, avoid obstacle and enermies and perhaps actually consist of the interaction of simple rules.

One's view

Private space 1　　　　　　　　Private space 2　　　　　　　　Private space 3

높낮이를 다양하게 하여 앉고 싶은 곳에 앉아서 쉴 수 있다.

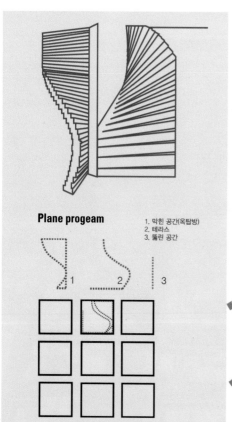

Plane progeam

1. 막힌 공간(옥탑방)
2. 테라스
3. 뚫린 공간

FUNCION PROCESS 1

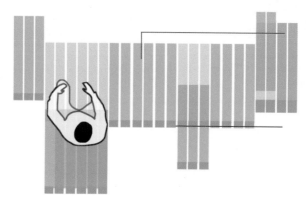

1. 높낮이를 다양하게 설정하여 의자와 테이블로 다양하게 사용할 수 있다.

Adding function of adjusting height of a vertical blind can be used for various aim like chair or table.

2.버티컬의 움직임과 방향성을 이용하여 옥탑방들의 프라이버시를 보호한다.

Protect the privacy using vertical blind of movement anddirectivity.

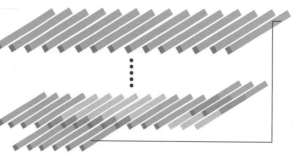

3. 앞, 뒤로 움직일 수 있고 위 아래 높낮이를 주어활용성을 높인다 .사선으로 구성하여 밖으로부터의 시선이 완벽히차단될 수 있다.

To enhance application level, make vertical blind move forward or backward and add function of adjusting height of a vertical slide. Making oblique vertical blind line can block glance from outside perfectly.

FUNCION PROCESS 2

눕다　　　기대다　　　앉다

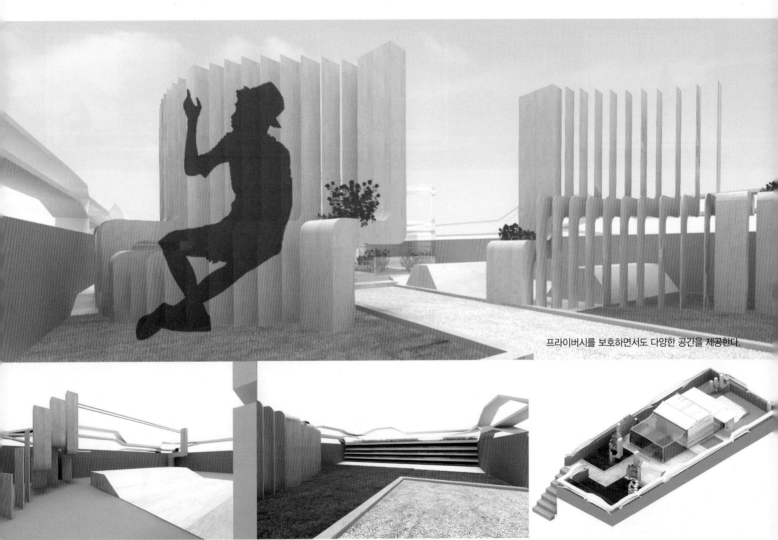

프라이버시를 보호하면서도 다양한 공간을 제공한다.

It is convenient to seat because of adding function of adjusting height of a vertical blind Also making vertical blind move forward or backward can make sement of clear (to secure the sight) and segment of block (to protect privacy).

View inside the bed room

침대가 뒤로 밀리면 아래 숨겨진 수납장이 드러나 수납을 할 수 있도록 설정하였다.

Space is formed

평범한 침대이지만 좁은 옥탑방의 공간활용을 높이기 위해 침대가 앞뒤로 움직이도록 하였다.

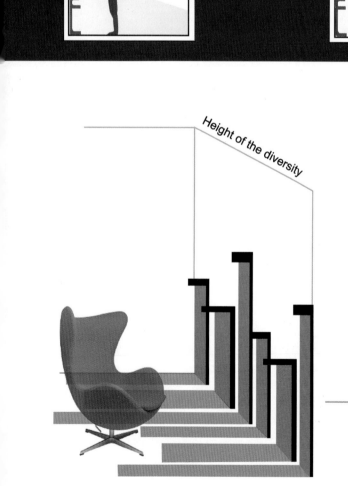

Height of the diversity

자급자족적 생활을 위해 옥상에 온실과 화단을

Utility Shelf

침실 옆 빈공간에 각각 높이차이가 있는 선반들을 두어 수납공간으로 활용할 수 있도록 하였다.

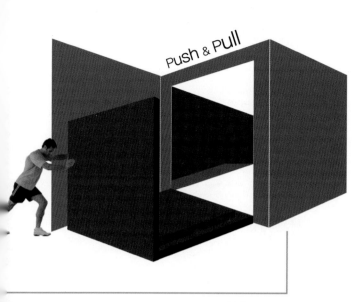

Push & Pull

Kitchen window

침실과 주방 사이에 공간을 두어 서로 통할 수 있도록 하였다. 또한 밖과 주방사이에 유리문을 내어 밖과 이어지도록 하였다.

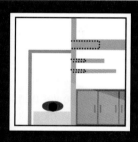

온실과 내부사이에 창을 두어 내부에서 쉽게
외부 온실을 볼 수 있도록 하였다.

ake greenhouse and flower garden on
e rooftop for self - sufficing life.

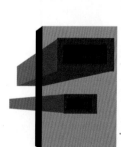

itchen window

과 주방사이에 유리문을 두어 안에서도 쉽게 자연 채광을 받을수있도록 하였다.

Push & pull

건물 외피사이의 틈에 창을 두어 내부로 자연채광이 들어오도록 하
였고, 침실 쪽의 수납벽이 주방과 연결되어 주방에서는 선반으로 사
용할 수 있도록 하였다.

ROOF

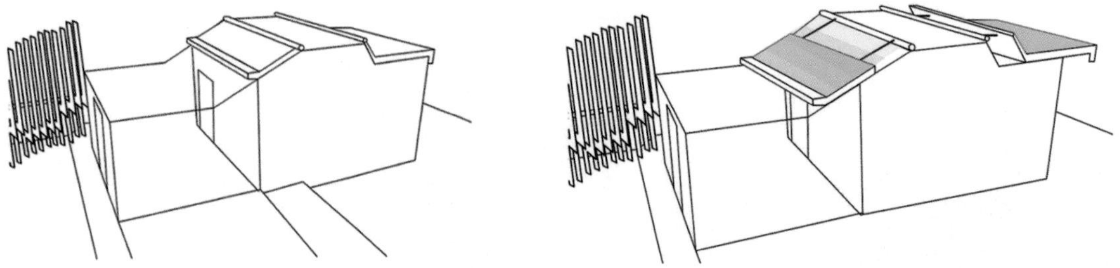

EXPLODED VIEW

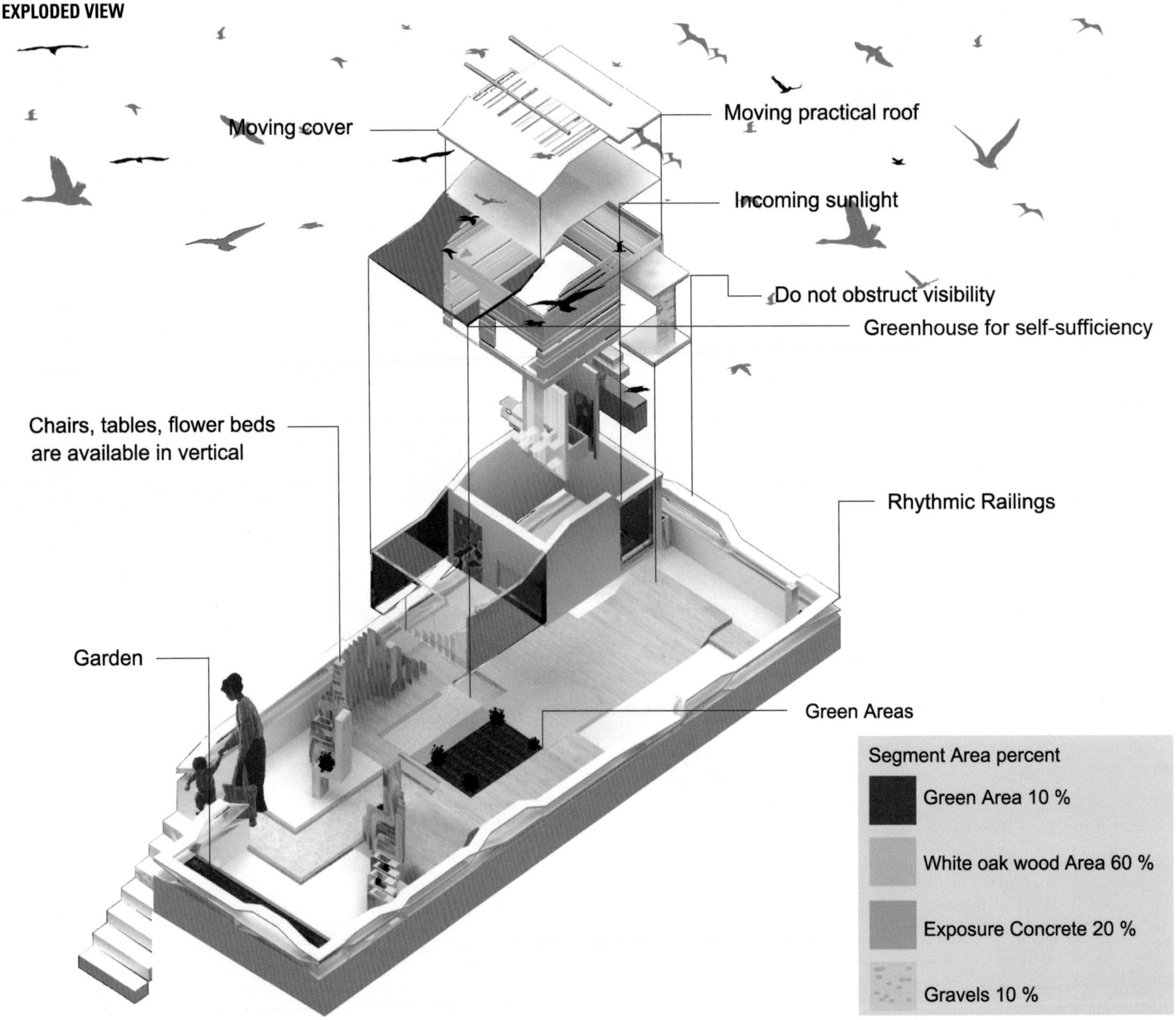

Moving cover

Moving practical roof

Incoming sunlight

Do not obstruct visibility

Greenhouse for self-sufficiency

Chairs, tables, flower beds are available in vertical

Rhythmic Railings

Garden

Green Areas

Segment Area percent

Green Area 10 %

White oak wood Area 60 %

Exposure Concrete 20 %

Gravels 10 %

Outside

Push & Pull

Outside Garden

Vertical

높이가 일정하고 빽빽하게 밀집되어 있는 옥탑방의 특징으로 인해 사생활이 침해되기 쉬웠던 문제를 버티컬의 개념을 도입하여 이를 해결하고 자급자족적인 생활을 할 수 있도록 다양한 기능을 구성한다.

Using vertical blind to solve problem that is ease to infringe privacy life and is caused by attribute that height is constant and is densely arranged. Beside vertical blind affords to lead self-sufficing life for people who live in rooftop house.

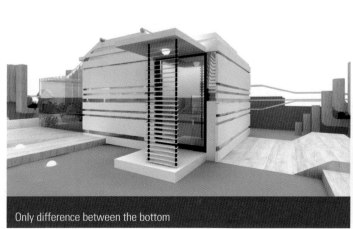

Only difference between the bottom

Long, thin window

2011
INTERNATIONAL DESIGN EXCHANGE PROJECT

The 2nd Hidden Space Project
URBAN ROOFTOPS

LONDON Metropolitan University
Sir John Cass Department of Art, Media and Design
Interior Design and Technology BA (Hons)
UK

DONGYANG MIRAE University
Department of Interior Design
Rep. of KOREA

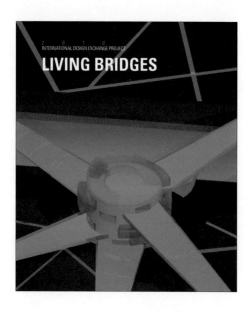

FHHS2010 - *The 1st Hidden Space Project* **"LIVING BRIDGES"**

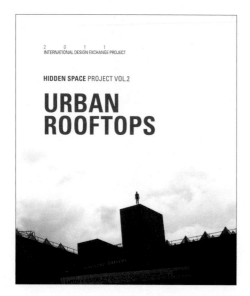

FHHS2011 - *The 2nd Hidden Space Project* **"URBAN ROOFTOPS"**

*FHHS (Future Horizon for Hidden Space)

2010년 [LIVING BRIDGES]주제로 영국 런던 메트로폴리탄대학과 한국의 경원대학교, 한국의 동양미래대학이 공동으로 런던 아키텍 페스티벌 참가하면서
시작된 **FHHS**는 우리시대 숨겨진 공간들을 10년간 10개의 주제로 교류전을 기획, 2회까지 진행했으며, 이후 주제를 현재 기획중이다.
올해는 2010년에 이어서 [URBAN ROOFTOPS]을 주제로 2번째 작품집이 발간되며, 또 7월중 런던에서 전시를 준비하고 있다.
FHHS는 이후 계속해서 우리주변의 숨겨진 공간들을 찾아 열정과 지혜, 그리고 성찰의 노정(路程) 을 계속해 나갈 예정이다.

2011

INTERNATIONAL DESIGN EXCHANGE PROJECT

The 2nd Hidden Space Project
URBAN ROOFTOPS

2011년 7월 1일 1쇄 발행

Published By 2011 URBAN ROOFTOPS Project
(Kaye Newman / Park, Young Tae / Ahn, Sung Hee)

Total Design

Park, Young Tae / Kim, Suk Young | DONGYANG MIRAE University

만든이 | 박영태, 김석영

편집,기획, 표지 디자인 | 박영태

Assistant

Kang, Ji Hee / Lee, Seo Young / Hwang, Eun Byeol | DONGYANG MIRAE University

편집 보조 | 강지희, 이서영, 황은별 – 동양미래대학 실내디자인과

발행처 | 도서출판 조경
인쇄 | 백산인쇄
출력 | 한결그래픽스

정가 | 30,000원